SpringerBriefs in Animal Sciences

More information about this series at http://www.springer.com/series/10153

David Steve Jacobs · Anna Bastian

Predator–Prey Interactions: Co-evolution between Bats and their Prey

 Springer

David Steve Jacobs
Animal Evolution and Systematics Group
Department of Biological Sciences
University of Cape Town
Cape Town
South Africa

Anna Bastian
Department of Biological Sciences
University of Cape Town
Cape Town
South Africa

ISSN 2211-7504   ISSN 2211-7512   (electronic)
SpringerBriefs in Animal Sciences
ISBN 978-3-319-32490-6   ISBN 978-3-319-32492-0   (eBook)
DOI 10.1007/978-3-319-32492-0

Library of Congress Control Number: 2016936429

Printed on acid-free paper

This Springer imprint is published by Springer Nature
The registered company is Springer International Publishing AG
The registered company address is: Gewerbestrasse 11, 6330 Cham, Switzerland

Preface

This book is aimed primarily at all scholars of evolution but particularly at those interested in predator–prey interactions. In a nutshell, the objective of this book is to explore whether bats and their prey have co-evolved. To this end, we review the biology of bats and their prey that is pertinent to the topic. The reviews here are therefore not as comprehensive as previous reviews with a more narrow focus on specific cases, but these are indicated throughout this book for those that are interested in greater detail. We aimed to provide a condensed yet comprehensive revision of bat–prey interaction which considers whether these interactions display the characteristics of specificity and reciprocity and thus represent cases of co-evolution.

Cape Town, South Africa

David Steve Jacobs
Anna Bastian

Acknowledgements

This book is based on research that has been supported in part by the University of Cape Town's Research Committee and the South African Department of Science and Technology's South African Research Chair Initiative administered by the National Research Foundation as well as a Postdoctoral Research Fellowship awarded by the South African National Research Foundation. We would like to thank Michael D. Greenfield and an anonymous reviewer for their constructive comments which improved the manuscript considerably, and last, but certainly not least, to those erudite researchers of bat, insect, and frog biology whose genius uncovered so many amazing, instructive, and entertaining phenomena that made this book enjoyable to write.

Contents

Chapter 1
Co-evolution

Abstract The precision of bat echolocation has continuously fascinated scientists since the eighteenth century. It enables bats to find prey as small as 0.05–0.2 mm in complete darkness and even to determine the kind of prey from the echo of its own call reflected off the prey. Besides active acoustic detection, some bat species also use passive listening for sounds generated by the prey itself, for example frog mating calls or the rustling sounds of crawling insects. Bats are therefore well adapted for hunting and can eat up to 25% of their body weight in insects each night, exerting much selection pressure on their prey. In response to this pressure, bat prey has evolved a range of defences both auditory and non-auditory. Frogs, for example, can listen for bat echolocation and cease their calling to female frogs. The evolution of audition in insects enables them to detect bats before the bats detect them, allowing insects to take evasive action. This interaction between predator and prey may represent a system of co-evolution. Here, we define co-evolution as a process in which the evolution of traits in the predator is in direct response to the evolution of traits in the prey which in turn evolved in direct response to the traits of the predator and so on. Co-evolution is thus an iterative process of both reciprocal (each lineage responds to the other) and specific (evolution of a trait in one lineage responds specifically to a trait in the other) adaptations.

1.1 Introduction

Echolocation, the detection of objects by means of echoes (Griffin 1944), is the only active system of orientation in the natural world. It is considered an active process because the acoustic signal upon which it is based is produced by the animal as opposed to passive systems which rely on external signals (e.g. light for vision and chemicals for smell) to gather information about the environment (Schnitzler and Henson 1980; Fenton 2001). Echolocation has evolved in animals that operate in low light conditions (e.g. cave swiftlets and marine mammals such as dolphins and whales) or in complete darkness (many species of bats). Bat echolocation is the

© The Author(s) 2016
D.S. Jacobs and A. Bastian, *Predator–Prey Interactions: Co-evolution between Bats and their Prey*, SpringerBriefs in Animal Sciences,
DOI 10.1007/978-3-319-32492-0_1

most sophisticated form of echolocation and allows bats to execute truly remarkable feats of orientation and prey capture.

Using echolocation, bats are able to detect the position of small insects over distances of several metres to within a few millimetres of accuracy (Simmons and Grinnell 1988) in complete darkness. Echolocation allows them to detect objects as small as 0.05–0.2 mm in diameter at distances of 1–2 m (Schnitzler and Henson 1980). Bat species that use specialized components of the echo, called acoustic glints (Chap. 2; Fig. 1.1), reflected off the beating wings of flying insects, can use changes in these glints caused by species-specific parameters encoded in the echo (e.g. wing beat rate) that are independent of the angular orientation of the prey relative to the bat. This allows them to distinguish one kind of insect prey from another (Von der Emde 1988; Jones 1990). They are also able to do so using other components of the echo generated by flying prey even when the wing beat rates are the same (Von der Emde 1988).

These abilities make bats extremely successful hunters. The small insectivorous species in the genus *Myotis* can catch and eat seven insects per minute or approximately 500 insects per hour (Gould 1955, 1959; Anthony and Kunz 1977) and consume up to 25% of their body weight in insects each night of foraging (Kunz 1974).

Some echolocating bats eat prey other than insects, and many of them prey on other vertebrates, e.g. lizards, other bats, fish, and frogs using a combination of echolocation and other auditory cues to detect these prey. For example, the fringe-lipped bat, *Trachops cirrhosis*, is very adept at catching frogs using both active echolocation (to detect movement of the vocal sac or the ripples in the water caused by the vibrating sac; Halfwerk et al. 2014) and passive listening to detect frogs. The latter involves bats using the mating calls of male frogs to detect and

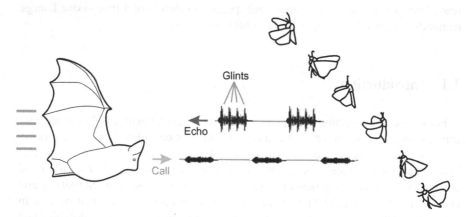

Fig. 1.1 Acoustic glints generated from the flapping wings of insects. Doppler shift compensation results in echoes of constant frequency from the background and stationary non-target objects. Changes in the amplitude and frequency of the echoes from the moving wings of the insect prey (in this case a moth) are superimposed on these constant frequencies and are called glints (color figure online)

capture these frogs during their mating season when male frogs call to attract female frogs (Chap. 6). Bats are so good at using such calls that they can discriminate between edible and poisonous frog species apparently using differences in the temporal components of the frogs' calls to do so (Ryan and Tuttle 1983; Chap. 6).

Bats are therefore extremely well adapted for hunting and exert enormous selection pressure on their prey which could promote the evolution of anti-bat defences. This is evidenced by the existence of adaptations that appear to diminish the risk of bat predation in both the vertebrate and invertebrate prey of bats. For example, frogs have evolved defensive behaviours that allow them to avoid detection by bats. Some frog species cease to call when bats are near (Page and Ryan 2008) or will only call within a chorus of frogs comprised of many frogs so that their probability of being taken by a bat is decreased (Ryan et al. 1981) through the dilution effect (Krebs and Davies 1993). Others have evolved mating calls that bats find difficult to localize (Chap. 6). Similarly, insects have evolved a range of defences that allow them to avoid being eaten by bats. By far, the most interesting of these are audition and ultrasonic clicks (Chap. 5). Many insects have evolved ears that they use to listen to the echolocation calls of bats. Such hearing acts as an early warning system allowing the insects to take evasive action. This works because most tympanate insects are able to hear the bat before the bats have detected them. This is so because insects are listening to the echolocation calls of the bat, but the bat has to wait for the softer returning echo of its call reflected off the insect to detect the insect. This means that tympanate insects can detect bats at greater distances than the bat can detect the insect (e.g. Goerlitz et al. 2010), provided the insects are paying attention. Insects with ears have a lower probability of being taken by a bat than atympanate insects (Griffin 1958; Chap. 4).

Several moth species have evolved ultrasonic sounds of their own in the form of a sequence of discrete clicks. These clicks are emitted when the bat is under attack and is very effective at forcing a bat to break off its attack (Chap. 4). Exactly how these clicks work is currently not known, but possible mechanisms include jamming—clicks interact with the bat's echolocation system in some, way interfering with the bats ability to track the moth in space and time. The clicks may also have an aposematic function warning the bat that the moth is unpalatable or the clicks may startle the bat in the same way that the eyespots on the hind wings of some moths startle visually oriented predators (Chap. 4).

Just as bat adaptations for hunting exerts selection pressure on their prey, the anti-bat defences of their prey may exert selection pressure on bats. If bats respond to such selection pressure, then bats and their prey may be locked into a special kind of evolutionary process in which changes in traits in the predator are influenced by the prey and vice versa. When two or more lineages interact in this way over several generations as host-parasite or predator–prey systems do, they could potentially each influence the evolution of the other. This could give rise to a special kind of evolution called co-evolution (Janzen 1980; Futuyma and Slatkin 1983; Futuyma 1986).

1.2   Co-evolution

Co-evolution is a special kind of evolution in which a trait in one lineage has evolved in response to a trait in another lineage and the trait in the second lineage has in turn evolved in response to the trait in the first (Janzen 1980; Futuyma 1986; Rothstein 1990). This results in reciprocal selection pressure being exerted on the interacting lineages. Co-evolution is likely to occur in systems where the same cue is used by the predator to detect prey and by the prey to detect that predator as, for example, in the use of echolocation calls by bats to detect moths through reflected echoes and the use of the bats' echolocation calls by moths to detect bats. However, the most important aspect of the definition of co-evolution that separates it from other forms of evolution is the requirement of reciprocity and specificity. Not only must traits in each lineage have evolved directly in response to traits in the other lineage (specificity) but also each lineage must in fact have responded to the evolution of the other (reciprocity). Thus, there must be an iterative process of both specificity and reciprocity (i.e. serial specificity and reciprocity) for the process to be co-evolution (Futuyma and Slatkin 1983) and excludes single responses that may or may not be reciprocal. For example, in some pollinator systems, the plant is under selection pressure to reduce the net costs of pollination by making it increasingly difficult for the pollinator to easily deplete all the flower's nectar while presenting its anthers in such a way that the pollinator has to make contact with the pollen. This facilitates pollen transfer to other flowers, while the flower retains sufficient nectar to attract other pollinators. The more pollinators that visit its flowers the better for the plant because it does not gamble all of its reproductive success on a single pollinator. Plants may not therefore co-evolve with any one pollinator, and the pollinators are also unlikely to evolve specializations to such plants. For example, many bat pollinated plants, such as the balsa tree, *Ochroma pyramidale*, are pollinated by several pollinators including bats and arboreal mammals (Fleming 1983), and the 'reciprocal adaptation syndromes' (von Helversen and Winter 2003) usually associated with co-evolution are not evident or represent a compromise for adaptation to a variety of pollinators (Waser et al. 1996). In contrast, many neotropical flowers (e.g. the bromeliad, *Werauhia gladioliflora*; Tschapka and von Helversen 2007) are specialized to be pollinated by bats in the subfamily the Glossophaginae. There are several plant characteristics that are adapted for pollen transfer to specific bats, but these plants also have characteristics that are adapted for pollination by this group of bats in general. These bat-pollination characteristics of plants are collectively called the chiropterophily syndrome (von Helversen and Winter 2003). Glossophagine bats appear to have responded to chiropterophily by evolving characteristics (e.g. elongated rostrums and tongues) that facilitate their access to nectar. This is taken to the extreme in the evolutionary interaction between the neotropical vine, *Centropogon nigricans*, and its exclusive pollinator, the tube-lipped nectar bat, *Anoura fistulata*. The vine has an elongated and narrow flower with the anthers at the entrance to the corolla and the nectar at the bottom to ensure pollen transfer as the bat laps up the nectar. In response to these tubular flowers, the bats have evolved a long rostrum with an extremely elongated tongue that can be extended to 85 mm which is 150% of its body length (Muchhala 2006). Thus, the reciprocal adaptive syndrome among interacting lineages, indicative of co-evolution, appears to be evident in these plants and their bat pollinators.

Another fundamental aspect of the definition of co-evolution as stated here is that it involves interacting lineages, i.e. non-mixing gene pools (Thompson 1982). Some authors (e.g. Krebs and Davies 1991; Holland and Rice 1999; Arnqvist and Rowe 2005; Greenfield 2016) regard sexual selection through female mate choice as a form of co-evolution. Although there may be an iterative process of both specificity and reciprocity between male traits and female choice for those traits, there are at least two glaring differences between sexual selection and co-evolution. Firstly, sexual selection does not involve the interaction between two or more lineages which is a key aspect of co-evolution. Secondly, the tandem evolution of female choice and exaggerated male traits requires that the alleles for both are genetically linked such that offspring would carry alleles for both the preference and the trait. Sons would then express the exaggerated trait, while daughters would express the preference for it (Fisher 1930). No such genetic linkage between co-evolving traits of interacting lineages is required in the definition of co-evolution. This makes sexual selection a very different process to co-evolution. Furthermore, both processes can result in similar consequences allowing at least the potential for similar consequences to erroneously conflate processes. For example, the evolution of sexual dimorphism in size and plumage in birds, usually the result of sexual selection, can also result from co-evolution between interacting lineages such as between host and parasite in avian brood parasitism (Krueger et al. 2007). The two processes are therefore best kept apart, and the use of co-evolution should be confined to instances which involve the interaction between different lineages.

The criteria of reciprocity and specificity are widely accepted, but some view these criteria, particular that of specificity, as too restrictive (Waters 2003) and advocate for a more relaxed definition which excludes specificity, as provided by Thompson (1994): 'a reciprocal evolutionary change in interacting species'. But in proposing this definition, Thompson (1982) makes it clear that he considers reciprocity and specificity among interacting lineages as key aspects of the definition of co-evolution. The use of reciprocal evolutionary changes in interacting species on its own to identify co-evolution could result in the conflation of disparate processes. For example, in predator–prey interactions, it is feasible that only one lineage, probably the prey, responds to the interaction, whereas the evolution of the other lineage could be a response to the habitat. This may be the case in the interaction between bats and their prey in which the evolutionary changes in prey hearing, for example, are the result of selection pressure from predation, but evolutionary changes in bat echolocation may instead be in response to the acoustic challenges of different habitats or different tasks (Chap. 5). Such a process would be very different to the process of co-evolution as it is widely accepted. It is essential that definitions are precise, even if exclusive. Inclusive but imprecise definitions are unlikely to be effective in our search for clarity and understanding of the myriad evolutionary processes responsible for biodiversity. Rarity of a process is not a sound reason for relaxing its definition. Although nature is organized along continua, exclusive definitions may assign processes that grade into one another into discrete categories, artificially separating them. When taking the first steps to understanding, exclusive definitions will prevent us from erroneously conflating processes that are in reality separate and different. If they are different parts of the same process, then once each is understood, we can better see the linkages between them.

Co-evolution does not preclude the traits being used in other contexts after they originated, only that their origin resulted from reciprocal and specific selection pressure. However, co-option for other functions could increase the difficulty in accumulating evidence for co-evolution because co-opted functions can mask the original function of the trait.

Besides, co-opted functions' interaction among several lineages may also make the demonstration of co-evolution difficult fuelling the perception that co-evolution is rare. Most lineages interact with several other lineages, and all of these may influence their evolution. For example, several bat species may interact with many insect species making it difficult to determine whether any defensive traits among the prey have co-evolved with offensive traits among the bat predators. If co-evolution is involved, this would be diffuse rather than pairwise co-evolution in which responses have a stepwise, reciprocal, and specific nature. Diffuse co-evolution is difficult to demonstrate because prey defences may be an accumulation of responses to several bat predators. Responding to many lineages may weaken the response to any one lineage making such responses difficult to identify. Nevertheless, diffuse co-evolution is also based on reciprocal responses to specific traits albeit in multiple lineages. However, pairwise interactions are more amenable to the identification of the specificity and reciprocity required by co-evolution (Rothstein 1990). Such systems include many mutualistic relationships, some parasite–host associations (Futuyma and Slatkin 1983), some pollination systems, and some predator–prey interactions. Among parasite–host associations, one of the most cited examples of co-evolution is avian brood parasitism.

Interspecific brood parasitism is a form of breeding in which individuals from one lineage or a few lineages (the parasites) lay their eggs in the nest of another or several other species (the hosts) duping the hosts into rearing the offspring of the parasites. This system is most prevalent in birds and hymenopterans (Payne 1977; Davies et al. 1989; Wcislo 1989) but is also found in fish (Feeney et al. 2012). A parasitic egg or hatchling in the nest of the host is extremely costly to the host because it may result in partial or complete destruction of the host's brood as well as the wastage of resources on raising the parasitic hatchling (Feeney et al. 2014a). On the other hand, it is extremely beneficial to the parasite because if successful the parasite has reproduced without the costs in time and energy involved in nest building, incubating eggs and rearing hatchlings. The stakes are high, and both host and parasite have much invested in its outcome each undergoing adaptation and counter-adaptation to overcome the defences and offensive strategies of the other to keep ahead in the race. An evolutionary arms race therefore adequately describes the process.

For example, evolutionary response by the host to prevent the parasite from laying its eggs in the host's nest, so-called 'frontline counter-adaptations' (Feeney et al. 2014a), takes the form of the hosts evolving behaviour that prevents the parasite from laying their eggs in the host's nest. This ranges from physically attacking or mobbing the parasites (Gloag et al. 2013) to less agonistic behaviours such as spending more time on the nest, thereby decreasing the opportunities for parasitism or simply physically blocking the parasite from laying its eggs (Gill and Sealy 2004; Canestrari et al. 2009).

The counter-adaptations by adult parasites to circumvent the host's defences are based largely on deceptive or cryptic adaptations, involving plumage and behaviour that prevent the host from recognizing the parasite (Feeney et al. 2014b). These adaptations may include cryptic plumage (Krueger et al. 2007), plumages that mimic predators (Gluckman and Mundy 2013), or plumage polymorphisms (Honza et al. 2006; Thorogood and Davies 2012, Trnka and Grim 2013) that include rare plumages that prevent host recognition of the parasite. Some parasites have also evolved cryptic behaviours, e.g. observing host nests from hidden perches (Honza et al. 2002) or attempting to lay eggs in host nests when hosts are absent (Davies 2000).

There is this much evidence of adaptation and counter-adaptation in all stages of the breeding interaction between avian hosts and their brood parasites providing clear examples of co-evolution (reviewed by Feeney et al. 2012, 2014b; Soler 2014). The other system most often cited as an example of co-evolution is the predator–prey interaction between bats and their tympanate insect prey, particularly moths.

Co-evolution between bats and insects was probably initiated by bats evolving echolocation which represented an offensive weapon that allowed bats to successfully hunt insects in complete darkness. Insects supposedly responded by evolving an effective defence against this weapon in the form of ears, which allowed them to hear bat echolocation, take evasive action and so avoid being eaten. Furthermore, some moths, already equipped with ears, have also evolved ultrasonic clicks of their own that are used defensively during a bat attack. Some bats in their turn supposedly improved their offensive weaponry by evolving stealth echolocation, i.e. echolocation at frequencies and intensities that are inaudible to moths (bats with such echolocation characteristics are also called 'whispering' bats (Griffin 1958; Jakobsen et al. 2013).

Stealth echolocation in bats can only be a co-evolved response if the selection pressure responsible for the evolution of stealth echolocation was exerted by insect hearing; that is, stealth echolocation evolved because it made bats less audible to insects and therefore more successful at capturing insect prey, and for no other reason. This does not preclude the use of stealth echolocation in contexts other than preying on insects after it originated, only that the reason for its origin was that it made bats less audible to insects. Similarly, insect hearing and the ultrasonic clicks of moths must have evolved because it provided them with an early warning system or rendered bat echolocation ineffective, respectively. In other words, insect defences originated in the context of bat predation because it increased the probability of insects surviving a bat attack. Once it originated, it could then be co-opted for some other purpose. Such co-option increases the difficulty in accumulating evidence for co-evolution because co-opted functions can mask the original function of the trait. A further difficulty in recognizing co-evolution between bats and their eared arthropod or vertebrate prey (i.e. frogs) is that in some insects (e.g. crickets and katydids; Gu et al. 2012; Strauß and Stumpner 2015) and frogs (Wells 2007), hearing evolved before bats and evolved as part of mating system in which acoustic signals are used in territorial defence, mate attraction, and mate recognition. So in these animals, the acoustic avoidance of bat predation was a co-opted function on an auditory system that evolved for other reasons. However, if it can be shown that at least some derived characteristics of the auditory systems of these animals were the result of a response to bat predation and that bats in turn responded to these characteristics, then a case can nevertheless be made for co-evolution.

We focus on the interaction between bats and their tympanate insect prey as a putative example of predator–prey co-evolution because it has been cited as such and is by far the most researched interaction involving bats in this context. This is so probably because of the large numbers of species of both tympanate insects and echolocating bats. We consider the interaction with other vertebrates as a means of illustrating the circumstances under which co-evolution is likely to emerge. We thus proceed by describing the adaptations of bat echolocation for prey detection and capture (Chap. 2) and the non-auditory (Chap. 3) and auditory (Chap. 4) defences of their insect prey. In Chap. 5, we consider whether bats and their insect prey have co-evolved. We focus on the interaction between bats and moths because this system has received the most attention and because, in addition to audition, moths have evolved the unique defence of using ultrasound, in the form of clicks, to counter bat echolocation. Analyses of moth clicks have opened a rich field of research on the function of ultrasonic signals in predator–prey interactions. It has the potential of improving our understanding of the evolution of acoustic signals in general including the underlying neurophysiology and its link to behavioural responses as well as the ecology of such responses. This is especially so within the context of evolutionary trade-offs, in this case, between predator evasion and reproduction (see Chap. 5). Finally, we use the unique system of bats and frogs in which bats rely mainly on passive hearing to hunt frogs to elucidate the circumstance under which co-evolution may emerge (Chap. 6). In this system, instead of a predator trait (i.e. bat echolocation) mediating the interaction between predator and prey, it is the prey's reproductive signals (viz. the frogs' mating song) that mediate the interaction between the two groups.

Although specialized carnivory has evolved independently in bats at least six times and carnivorous bats show adaptations for carnivory, including increased bite force at wide gape angles (Santana and Cheung 2016), the interaction between carnivorous bats and their vertebrate terrestrial prey has not been investigated in the context of co-evolution. We do not therefore consider them here (but see Chap. 7). The interaction between bats as pollinators and the plants they visit is not a predator–prey system and not therefore covered here, but this system nevertheless represents a fruitful area of research on co-evolution (e.g. Muchhala 2006). In the final chapter (Chap. 7), we provide a synthesis of the topic and describe some of the future research that is required to advance our understanding of co-evolution in particular and evolution in general.

References

Anthony ELP, Kunz TH (1977) Feeding strategies of the Little Brown bat, *Myotis lucifugus*, in Southern New Hampshire. Ecology 58:775–786

Arnqvist G, Rowe L (2005) Sexual conflict. Princeton University Press, New Jersey

Canestrari D, Marcos JM, Baglione V (2009) Cooperative breeding in carrion crows reduces the rate of brood parasitism by great spotted cuckoos. Anim Behav 77(5):1337–1344

Davies NB (2000) Cuckoos, cowbirds and other cheats, 1st edn. T. & A. D. Poyser Ltd., London

Davies NB, Bourke AFG, Brooke MD (1989) Cuckoos and parasitic ants—interspecific brood parasitism as an evolutionary arms-race. Trends Ecol Evol 4(9):274–278

Feeney WE, Welbergen JA, Langmore NE (2012) The frontline of avian brood parasite-host co-evolution. Anim Behav 84(1):3–12

Feeney WE, Stoddard MC, Kilner RM, Langmore NE (2014a) "Jack-of-all-trades" egg mimicry in the brood parasitic Horsfield's bronze-cuckoo? Behav Ecol 25(6):1365–1373

Feeney WE, Welbergen JA, Langmore NE (2014b) Advances in the study of co-evolution between avian brood parasites and their hosts. In: Futuyma DJ (ed) Annual review of ecology, evolution, and systematics, vol 45. Annual Reviews, Palo Alto, pp 227–246

Fenton MB (2001) Bats: revised edition. Checkmark Books, New York

Fisher RA (1930) The genetical theory of natural selection: a complete, variorum edn. Oxford University Press, Oxford

Fleming TH (1983) Carollia perspicillata. In: Janzen DH (ed) Costa Rican natural history. University of Chicago Press, Chicago

Futuyma DJ (1986) Evolutionary biology. Sinauer Associates Inc, Sunderland, Massachusetts

Futuyma DJ, Slatkin M (1983) Introduction. In: Co-evolution. Sinauer Associates Inc., Sunderland, Massachusetts

Gill SA, Sealy SG (2004) Functional reference in an alarm signal given during nest defence: seet calls of yellow warblers denote brood-parasitic brown-headed cowbirds. Behav Ecol Sociobiol 56(1):71–80

Gloag R, Fiorini VD, Reboreda JC, Kacelnik A (2013) The wages of violence: mobbing by mockingbirds as a frontline defence against brood-parasitic cowbirds. Anim Behav 86 (5):1023–1029

Gluckman TL, Mundy NI (2013) Cuckoos in raptors' clothing: barred plumage illuminates a fundamental principle of Batesian mimicry. Anim Behav 86(6):1165–1181

Goerlitz HR, ter Hofstede HM, Zeale MRK, Jones G, Holderied MW (2010) An aerial-hawking bat uses stealth echolocation to counter moth hearing. Curr Biol 20(17):1568–1572

Gould E (1955) The feeding efficiency of insectivorous bats. J Mammal 36:399–407

Gould E (1959) Further studies on the feeding efficiency of bats. J Mammal 40:149–150

Greenfield MD (2016) Evolution of acoustic communication in insects. In: Pollack GS, Mason AC, Popper AN, Fay RR (eds) Insect hearing. Springer Handbook of Auditory Research, vol 55, pp 17–47

Griffin DR (1944) Echolocation by blind men, bats and radar. Science 100(2609):589–590. doi:10. 1126/science.100.2609.589

Griffin DR (1958) Listening in the dark: the acoustic orientation of bats and men. Yale University Press, New Haven

Gu JJ, Montealegre-Z F, Robert D, Engel MS, Qiao GX, Ren D (2012) Wing stridulation in a Jurassic katydid (Insecta, Orthoptera) produced low-pitched musical calls to attract females. Proc Natl Acad Sci USA 109(10):3868–3873

Halfwerk W, Dixon MM, Ottens KJ, Taylor RC, Ryan MJ, Page RA, Jones PL (2014) Risks of multimodal signaling: bat predators attend to dynamic motion in frog sexual displays. J Exp Biol 217(17):3038–3044

Holland B, Rice WR (1999) Experimental removal of sexual selection reverses intersexual antagonistic co-evolution and removes a reproductive load. Proc Nat Acad Sci USA 96 (9):5083–5088

Honza M, Taborsky B, Taborsky M, Teuschl Y, Vogl W, Moksnes A, Roskaft E (2002) Behaviour of female common cuckoos, *Cuculus canorus*, in the vicinity of host nests before and during egg laying: a radiotelemetry study. Anim Behav 64:861–868

Honza M, Sicha V, Prochazka P, Lezalova R (2006) Host nest defense against a color-dimorphic brood parasite: great reed warblers (*Acrocephalus arundinaceus*) versus common cuckoos (*Cuculus canorus*). J Ornithol 147(4):629–637

Jakobsen L, Brinkløv S, Surlykke A (2013) Intensity and directionality of bat echolocation signals. Frontiers Physiol 4:1–9

Janzen DH (1980) When is it co-evolution? Evolution 34(3):611–612

Jones G (1990) Prey selection by the Greater horseshoe bat (*Rhinolophus ferrumequinum*): optimal foraging by echolocation? J Anim Ecol 59(2):587–602

Krebs JR, Davies NB (1991) Behavioural ecology: an evolutionary approach, 3rd edn. Blackwell Science Ltd., Oxford

Krebs JR, Davies NB (1993) An introduction to behavioural ecology, 3rd edn. Blackwell Scientific Publications, Oxford

Krueger O, Davies NB, Sorenson MD (2007) The evolution of sexual dimorphism in parasitic cuckoos: sexual selection or co-evolution? Proc Roy Soc Lond B Biol Sci 274(1617): 1553–1560

Kunz TH (1974) Feeding ecology of a temperate insectivorous bat (*Myotis velifer*). Ecology 55:693–711

Muchhala N (2006) Nectar bat stows huge tongue in its rib cage. Nature 444:701–702

Page RA, Ryan MJ (2008) The effect of signal complexity on localization performance in bats that localize frog calls. Anim Behav 76:761–769

Payne RB (1977) The ecology of brood parasitism in birds. Annu Rev Ecol Syst 8:1–28

Rothstein SI (1990) A model system for co-evolution: Avian brood parasitism. Annu Rev Ecol Syst 21:481–508

Ryan MJ, Tuttle MD (1983) The ability of the frog-eating bat to discriminate among novel and potentially poisonous frog species using acoustic cues. Anim Behav 31:827–833

Ryan MJ, Tuttle MD, Taft LK (1981) The costs and benefits of frog chorusing behavior. Behav Ecol Sociobiol 8(4):273–278

Santana SE, Cheung E (2016) Go big or go fish: morphological specializations in carnivorous bats. Proc Roy Soc B 283:20160615. doi:10.1098/rspb.2016.0615

Schnitzler H-U, Henson OW (1980) Performance of airborne animal sonar systems: I. Microchiroptera. In: Busnel R-G, Fish JF (eds) Animal sonar systems, Plenum Press, New York, pp 109–181

Simmons JA, Grinnell AD (1988) The performance of echolocation: acoustic images perceived by echolocation bats. In: Nachtigall PE, Moore PWB (eds) Animal sonar: processes and performance. Plenum Press, New York, pp 353–385

Soler M (2014) Long-term co-evolution between avian brood parasites and their hosts. Biol Rev 89(3):688–704

Strauß J, Stumpner A (2015) Selective forces on origin, adaptation and reduction of tympanal ears in insects. J Comp Physiol A Neuroethology Sens Neural Behav Physiol 201(1):155–169

Thompson JN (1982) Interaction and co-evolution. Wiley, New York

Thompson JN (1994) The co-evolutionary process. University of Chicago Press, Chicago

Thorogood R, Davies NB (2012) Cuckoos combat socially transmitted defenses of reed warbler hosts with a plumage polymorphism. Science 337(6094):578–580

Trnka A, Grim T (2013) Color plumage polymorphism and predator mimicry in brood parasites. Frontiers Zool 10:1–10

Tschapka M, von Helversen O (2007) Phenology, nectar production and visitation behaviour of bats on the flowers of the bromeliad *Werauhia gladioliflora* in a Costa Rican lowland rain forest. J Trop Ecol 23:385–395

Von der Emde G (1988) Greater horseshoe bats learn to discriminate simulated echoes of the insects fluttering with different wing beats. In: Nachtigall PE, Moore PWB (eds) Animal sonar: processes and performance. Plenum Press, New York, pp 495–500

von Helversen O, Winter Y (2003) Glossophagine bats and the their flowers: costs and benefits for plants and pollinators. In: Kunz TH, Fenton MB (eds) Bat ecology. Univesity of Chicago Press, Chicago, pp 346–397

Waser NM, Chittka L, Price MV, Williams N, Ollerton J (1996) Generalization in pollination systems, and why it matters. Ecology 77:1043–1060

Waters DA (2003) Bats and moths: what is there left to learn? Physiol Entomol 28(4):237–250

Wcislo WT (1989) Behavioral environments and evolutionary change. Annu Rev Ecol Syst 20:137–169

Wells KD (2007) The ecology and behavior of amphibians. 1st edn. The University of Chicago Press, Chicago

Chapter 2
Bat Echolocation: Adaptations for Prey Detection and Capture

Abstract Bats have evolved a plethora of adaptations in response to the challenges of their diverse habitats and the physics of sound propagation. Such adaptations can confound investigations of adaptations that arise in response to prey. Here, we review the adaptations in the echolocation and foraging behaviour of bats. Bats use a variety of foraging modes including aerial hawking and gleaning. The main challenge to bats echolocating in clutter is increased resolution to detect small objects, be they insects or twigs, and overcoming the masking effects that result from the overlap of echoes from prey and the background. Low-duty-cycle echolocating bats that aerial hawk in clutter have evolved short, frequency-modulated calls with high bandwidth that increase resolution and minimizes masking effects. Bats that glean prey from the vegetation have similar adaptations but in addition use passive listening and/or 3-D flight to ensonify substrate-bound prey from different directions. High-duty-cycle echolocating bats have evolved Doppler-shift compensation which allows the detection of acoustic glints from the flapping wings of insects. Bats that aerial hawk in the open have to search large volumes of space efficiently and use narrowband, low-frequency (for decrease atmospheric attenuation) echolocation calls that maximize detection distance. The wings and echolocation of bats form an adaptive complex. Bats that forage in clutter where distances are short have short broad wings for slow, manoeuvrable flight. Bats that hunt in the open where detection distances are long have long narrow wings for fast, agile flight.

2.1 An Introduction to Echolocation

Echolocation is the location of objects in time and space using reflected sound (i.e. an echo) rather than the reflected light used in vision. The use of reflected sound means that echolocation can be used in complete darkness. The sophisticated echolocation of modern bats probably evolved in the context of the detection and capture of small insects on the wing in complete darkness (Griffin 1944). During echolocation, bats generate their echolocation pulses or calls using their larynx and

© The Author(s) 2016
D.S. Jacobs and A. Bastian, *Predator–Prey Interactions: Co-evolution between Bats and their Prey*, SpringerBriefs in Animal Sciences,
DOI 10.1007/978-3-319-32492-0_2

vocal chords, in much the same way humans do when we speak. The call travels through the air in the form of concentric sound pressure waves, and when these waves are reflected off objects, echoes are generated and these in turn travel back through the air from the object as another series of sound pressure waves that the bats' ears can detect. The bat's brain interprets differences between its emitted call and the returning echo to form a three-dimensional image of its surroundings in much the same we use our eyes and light to form three-dimensional images of our surroundings.

The echolocation systems used by bats are sophisticated and complex, but the manner in which echolocation functions can be understood through acoustics, the theory of sound. Acoustics has allowed the documentation of the diversity of echolocation types/systems among bats. In combination with flight, echolocation has allowed bats to colonize just about every conceivable nocturnal niche and adaptations to these niches have given rise to more than 1270 species of bats (Matthias et al. 2016). This has led to an enormous diversity in form and behaviour, making bats the most ecologically diverse group of mammals.

Although all bats fly, not all of them echolocate and about 184 species, commonly known as Flying foxes rely almost exclusively on vision. However, at least three species of Flying fox, *Rousettus aegyptiacus*, *R. amplexicaudatus* (Novick 1958; Griffin 1958; Holland et al. 2004; Yovel et al. 2011), and *R. leschenaulti* (Novick 1958; Raghuram et al. 2007; Fenton and Ratcliffe 2014) also use a rudimentary form of echolocation in which echolocation calls are generated by tongue clicks (lingual echolocation) rather than by the larynx (laryngeal echolocation). Although it has been suggested that other species of Flying fox may also use lingual echolocation (Funakoshi et al. 1995; Schoeman and Goodman 2012), this has not been confirmed. A third type of echo sensing, considered to be rudimentary because it does not allow the accurate estimation of object distance, involves the generation of clicks using the wings (Gould 1988; Boonman et al. 2014). Old World fruit bats are able to use echoes from these wing clicks to detect objects in complete darkness (Boonman et al. 2014).

2.1.1 Detection and Classification of Targets

During echolocation, a bat will typically emit a series of calls which will be reflected off objects as echoes. The bat hears the echoes with its ears and the bats brain can match the spectral and temporal properties of the emitted call with that of the echo to form a three-dimensional image of its surroundings in much the same way we use vision to do so. Calls of laryngeal echolocators can be emitted either through their mouths (mouth echolocators) or nostrils (nasal echolocators), depending on the species. The complex folds and flaps of skin around the nostrils of nasal echolocators and the shape of the pinnae and head of all bats are associated with the emission and reception of sound during echolocation (Pedersen 1993; Obrist et al. 1993; Goudy-Trainor and Freeman 2002; Zhuang and Müller 2006).

   Bats are able to use their echolocation to determine the size, shape, material, texture distance of an object and perhaps also the type of insect prey. Bats can also differentiate between echoes from a prey item and those from non-target objects such as the background substrate or vegetation (clutter echoes). All of this information is encoded in the complex temporal and spectral changes that the bat's call undergoes when reflected as an echo. The acoustics involved is very complex but can be simplified as follows. The size of an object is determined simply by the intensity of the echo, larger objects generate echoes of greater intensity than smaller objects. Objects smaller than one wavelength of the echolocation call will not generate any echo or only weak echoes and the bat would not be able to detect it. Echolocation calls are scattered by objects smaller than one wavelength. This is known as Rayleigh scattering (Young 1981; Pye 1993). Rayleigh scattering causes sound reflected by an object smaller than one wavelength of the sound to be scattered in all directions so that very little of it gets back to the bat. As a result, most echolocating bats have to use echolocation calls of short wavelength and therefore high frequency so that echoes are generated from small leaves and twigs (which the bat has to avoid) and from small food items (e.g. small insects) which the bat has to catch and eat. Although most bats echolocate between 20 and 50 kHz (Fullard 1987), the full range of call frequencies used by bats is 9–210 kHz (Fenton et al. 2004; Jacobs 2000).

   The speed of sound is constant at atmospheric conditions at which bats operate and bats can therefore determine the distance between it and an object by the time elapsed between when it emits a call and the echo returns from the object. Bats are capable of detecting very small time delays that allow them to differentiate between objects whose distances from the bat differ by as little as 1 mm (Neuweiler 1990). The relative intensity of sound received at each ear and the time delay between arrivals at the two ears provide information about whether the object is to the left or right of the bat, similar to the method first discovered in owls to passively locate prey using the sounds generated by the prey (Payne 1971). Similarly, interference patterns caused by echoes reflecting from the tragus, a flap of skin rising from the bottom margin of the external ear, allow the bat to determine whether the object is below or above the bat (Chiu and Moss 2007).

   Echolocation also allows bats to resolve one object from another or from the background and to obtain information about the texture (Schmidt 1988; Simon et al. 2014) of an object. They do so by using calls comprising different wavelengths of sound and by shifting the peak frequency (frequency of maximum energy) of their calls (Neuweiler 1990; Schnitzler and Kalko 2001).

   When two objects are close together, the echoes from each of them interfere with those from the other and will produce different patterns of sound waves than those generated by only one object. This allows the bat to discriminate objects that are less than 1 mm apart (Neuweiler 1990). Similarly, objects with very complex surfaces will reflect waves in different directions producing different interference patterns to that generated by objects with smoother surfaces allowing the bat to determine the texture of the object.

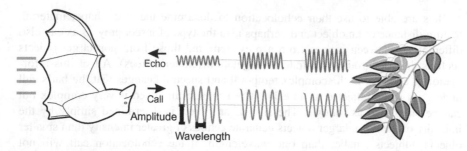

Fig. 2.1 Atmospheric attenuation and Doppler-shift compensation. **Atmospheric attenuation**: The emitted call (in *blue*) travels through the atmosphere in the form of sound pressure waves. As each pressure wave is formed, some of the energy in the sound is dissipated (indicated by the reduction in the amplitude of wave). When the sound is reflected off the object and travels back to the bat as an echo (in *red*), the echo is also attenuated. The echo that the bat hears is therefore much softer than the call that impinges on the object. **Doppler-shift compensation**: The bat emits its call (in *blue*) at a slightly lower frequency than the reference frequency (the frequency of its acoustic fovea—see text). Because of the relative motion of the bat, the echo from the object is shifted up in frequency as a result of the Doppler effect indicated by the shorter wavelengths in the echo in red. The bat compensates for the Doppler effect so that the echo always returns at the reference frequency (color figure online)

Atmospheric attenuation (Fig. 2.1) places limitations on the distance over which echolocation is effective and echolocation is therefore a short-range detection system operating over a few metres only. This is because bats have to use high-frequency sound to detect small objects such as insects and high-frequency sound is especially affected by atmospheric attenuation (Lawrence and Simmons 1982). Apart from frequency of the sound, atmospheric attenuation is also dependent on the temperature and humidity of the air. It is more pronounced for calls of higher frequency and increases as the temperature and humidity increases. All else being equal, bats that echolocate at the higher end of the frequency range used by bats should detect objects at shorter distances than bats that echolocate at the lower end of this range (Luo et al. 2013). However, it appears that bats can compensate for frequency-dependent atmospheric attenuation by increasing the intensity of their echolocation calls. Bat species with similar hunting methods and hunting in habitats from the edge of vegetation to open space away from vegetation or over water have similar detection distances. This is largely because bats echolocating at higher frequencies also use much higher call intensities so that they all have more or less the same detection ranges, i.e., increased intensity can counter frequency-dependent atmospheric attenuation (Surlykke and Kalko 2008).

Besides increasing the intensity of the call, bats are able to increase the distance over which their echolocation is effective by decreasing the band width of the call so that the energy of the call can be concentrated in a narrow range of frequencies or by decreasing the frequency of the call to minimize atmospheric attenuation. However, these strategies have trade-offs associated with them. Decreasing the bandwidth of the call decreases the resolving power of the call. Decreasing the frequency means that the bat will not be able to detect small objects. Thus, increasing detection range decreases resolution and vice versa. Different species of bats have evolved different ways of dealing with these trade-offs to optimize the effectiveness of echolocation resulting in many different kinds of calls consisting of different components.

Target size and texture, the wing-beat patterns of insects (Rhebergen et al. 2015), and movement of the vocal sac of frogs (Halfwerk et al. 2014) could allow bats to classify the type of prey they are tracking using their echolocation calls. However, some bats using high-duty-cycle echolocation (see section on duty cycle below) are also able to determine the type of insect prey they are tracking from the temporal and spectral information in the echoes reflected off their prey (von der Emde and Schnitzler 1990) and from the wing-beat frequency of the prey (von der Emde and Menne 1989). High-duty-cycle bats detect prey by the amplitude and frequency modulations (acoustic glints, Fig. 1.1) generated when the calls are reflected off the wings of the prey which are at different angles to the impinging echolocation beam during the insect's wing-beat cycle. These bats can accurately measure the wing-beat frequency of insects by measuring the rate at which the acoustic glints are generated. Since wing-beat frequencies are inversely correlated with the size of insects, it can be used as the first step in classification. However, these bats can also use the amplitude and frequency information in the glints and the echoes between glints (which are unique for each type of insect even if they have the same glint frequency) to determine the type of insect, irrespective of the orientation of the insect relative to the bat (von der Emde and Menne 1989; von der Emde and Schnitzler 1990).

2.1.2   Components of Calls

2.1.2.1   Frequency-Modulated Components (FM)

Bat calls are composed of various components of mainly two types, frequency-modulated (FM) and constant-frequency (CF) components but with a range of variations of these two types of components. The calls of most species of bats are dominated by FM components while those of the remainders (largely just three families of bats, Rhinolophidae, Rhinonycteridae, and Hipposideridae) are dominated by CF components. FM calls can be narrowband, with a low range of frequencies or broadband with a high range of frequencies. Narrowband FM signals are good for target detection but not for target localization and classification. Broadband FM calls are good for target localization and classification but because of atmospheric attenuation and the consequent short detection range are not good for target detection (Schnitzler and Kalko 2001).

Broad bandwidth FM calls allow exact target localization through accurate determination of the distance and angle to the target. Echoes from steep FM broadband calls activate the neuronal filters in the brain of the bat for only a short time, producing the precise time markers that allow the accurate determination of the time delays that encode the distance to the target (Goerlitz et al. 2010). In addition the large bandwidth activates more neuronal filters, improving the accuracy of both the distance (Moss and Schnitzler 1995) and angle determination with increasing bandwidth (Schnitzler and Kalko 2001).

Frequency-modulated signals of broad bandwidth also allow the determination of texture and depth structure of the target. Texture of the target can be determined from the differences in absorption of different frequencies while depth structure can be determined from interference patterns caused by the overlapping of multiple sound waves in the echoes from the target. Determination of these parameters could assist the bat in the classification of the target (Neuweiler 1989, 1990).

To optimize detection range and target localization most bats use a combination of narrowband and broadband FM calls switching from one to the other depending on the situation. Most species also use harmonics (Neuweiler 1984; Schnitzler and Kalko 2001) to broaden the bandwidth of their calls. Echolocation calls consist of the fundamental or 1st harmonic, and a series of harmonics of higher frequency called the 2nd, 3rd and 4th harmonics and so on. Not only are these harmonics important because they allow the bat to resolve smaller objects than they otherwise would have been able to, but they also help the bat to distinguish echoes from the target and echoes from non-target objects (i.e. clutter) in the background or to the periphery of the target (Bates et al. 2011).

Echoes from clutter can interfere with the bats processing of target echoes, especially when the two kinds of echoes arrive at the bats auditory system at the same time. This masking of target echoes by echoes from clutter (Fig. 2.2) makes it difficult for the bat to use time delays to determine distances to the target. The horizontal width of the echolocation beam is dependent on the frequency of the sound; the higher frequency of higher harmonics has a narrower beam than the lower frequency of the 1st harmonic and therefore generates fewer echoes and echoes of lower amplitude from the clutter than from the target. The target reflects echoes from both the 1st and the higher harmonics that are of higher amplitude than echoes from clutter allowing the bat to focus on echoes from the target. This is analogous to visual images of objects directly in front of us and those in our peripheral vision. The harmonic structure of echolocation calls also assists in isolating echoes from the target from background clutter because the higher-frequency sounds of higher harmonics attenuate more quickly in air. Echoes generated by the higher harmonics reflect more weakly off the more distant background than off the closer target. Echoes from the target therefore are of higher amplitude than echoes from the background allowing the bat to detect an insect, for example, against the background clutter (Bates et al. 2011).

Although it is possible that a bat could increase its bandwidth by calls sweeping from higher to lower frequencies instead of harmonics, the duration of such calls would be longer than calls with harmonics and would increase the potential of temporal overlap of background echoes with all or parts of the call, interfering with the bat's ability to process target echo time delays. The facility with which FM components can be used to resolve targets while avoiding call echo overlap when close to clutter makes them ideal for hunting in dense vegetation where resolution is

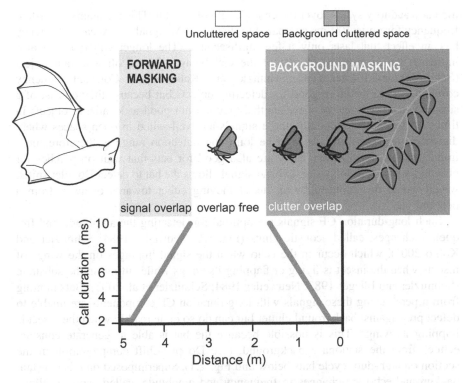

Fig. 2.2 Masking effects. **Forward masking** is the overlap of call and echo and occurs in the zone immediately in front of the bat (*grey area*). **Backward masking** is the overlap of target echo with echoes from the background (*green areas*). There is an overlap-free zone some distance from the bat and the vegetation. This zone increases in size (*orange lines*) as the bat decreases the duration of its calls. Adapted from Schnitzler and Kalko (2001) (color figure online)

of greater importance than detection distances (Schnitzler and Kalko 2001). The distance between bat and objects is small in cluttered situations in any case.

2.1.2.2   Constant-Frequency (CF) Components

Constant-frequency (CF) components of calls are components in which there is no change in frequency with time. In such components, the energy is concentrated into a single or narrowband of frequencies allowing the sound to travel through the atmosphere for greater distances, before atmospheric attenuation dissipates the energy, than that in FM components in which the energy is spread over several frequencies. Focusing all of the energy in a narrowband of frequencies also allows the CF signal to be of longer duration than FM signals. The operational range of CF components is therefore further than that of an FM signal because echoes returning within the narrow-frequency band are of higher amplitude and can be summed by the bat's auditory system over the entire length of the call. This maintains a constant frequency for up to 100 ms compared to that in a FM signal where each frequency has an effect that lasts only a few milliseconds. The longer call duration also increases the probability of part of the call being reflected off a distant object (Schnitzler and Flieger 1983; Schnitzler and Kalko 2001). Constant-frequency components are therefore good for detecting objects but because they consist of a single or narrow range of wavelengths, they are not good at localizing objects in time and space. Consequently, these signals are well-suited in open spaces where distances to objects are likely to be long and detection range is therefore more important than localization. They are also good for bats that wait on perches for prey to fly past. The long, narrowband signal allows the bat to detect Doppler shifts, which would be produced by an insect moving either towards or away from a perched bat.

Such long-duration CF signals are optimal for detecting the amplitude and frequency changes, called acoustic glints (Fig. 1.1; Neuweiler 1984; Schnitzler and Kalko 2001), which occur in the echo when the signal impinges on the wings of insects when the insect is flying or flapping its wings while sitting on the substrate (Schnitzler and Flieger 1983; Neuweiler 1984; Schnitzler et al. 2003). Bats hunting from a perch using these signals with long-duration CF components are unable to detect prey against background clutter but can do so quite readily when the insect is flapping its wings. This is possible because the bat is able to generate constant echoes from the stationary background, (see Doppler-shift compensation in the section on high-duty-cycle bats below and Fig. 2.1). Superimposed on this constant background echo are changes in frequency and amplitude, called acoustic glints, generated when the call is reflected by the fluttering insect wings (Schnitzler and Flieger 1983; Neuweiler 1984; Schnitzler and Kalko 2001). When the long-duration CF signal impinges upon a wing, the amplitude and frequency of the echo are dependent on the position of the wing and whether the wing is moving towards the bat or away from it, respectively. Remember that the strength (i.e. amplitude) of an echo is dependent on the size of the object generating it. When the insect wing is in an upright position or at the bottom of its arc and perpendicular to the impinging echolocation signal all of its surface area reflects the signal producing an echo of maximum amplitude. When the wing is halfway through its cycle only its edge is exposed to the impinging signal and a very weak echo is produced. Similarly when the insect wing closest to the bat is moving down towards the mid-point of its cycle, it is moving towards the bat and the Doppler effect shifts the frequency of the echo up relative to the frequency in the bat's signal. When the wing is moved past the mid-point of its cycle and towards its lowest point, it is moving away from the bat and the echo is Doppler shifted lower than the bat's signal. These changes in the amplitude and frequency in the echoes of a bat's call from flapping insect wings are perceived by the bat as amplitude and frequency glints against the constant echo from the background clutter collectively called acoustic glints (Schnitzler and Flieger 1983; Neuweiler 1984; Schnitzler and Kalko 2001), which allow the bats to detect flapping inset wings against the background clutter. Acoustic glints are

analogous to coloured lights of different brightness flashing on and off against a
background lit by a light of constant colour and intensity.

2.1.3   Duty Cycle

Bats emit their echolocation signals as discrete packets of sound or calls. These
calls are thus separated by silence during which the bat is emitting no sound. The
ratio of the duration of the call to the period of the call (call duration + time to the
next call) is the duty cycle (Fenton 1999) and is usually expressed as a percentage.

2.1.3.1   Low-Duty-Cycle (LDC) Bats

Low-duty-cycle (LDC) bats have call durations that are short relative to the period
of the call. The reason for this is the way in which these bats avoid being deafened
by their own calls. Atmospheric attenuation at the high frequency at which bats
operate means that bats have to emit their calls at very high intensities (measured as
110–135 dB at 10 cm in front of the bat) to have a functional range of detection.
However, the inner ears of the bat have to be sensitive enough to pick up the
faintest of echoes. Such a sensitive ear is likely to be damaged by sounds great than
90 dB emitted only millimetres from the ear. To avoid damage to the inner ear,
LDC bats switch off their ears. In the middle ear of bats, the three small ossicles are
linked in tandem and transmit sound from the ear drum to the cochlear (or inner
ear). Attached to the ossicles are small muscles. Just before the bat emits its
echolocation call, these muscles contract and disrupt the linkage between the three
ossicles preventing the transmission of sound to the cochlear, thereby protecting the
sensitive inner ear. The bat then emits the echolocation call. After emission of the
call, it relaxes the muscles and restores the link between the ossicles allowing sound
to be transmitted to the cochlear. The bat thus separates the emitted call and
returning echo in time and has to wait until the echo from one call returns before
emitting the next call so that the echo from one call does not return while the ear is
functionally deaf during the emission of the next call. This is why there is a
relatively long period of silence between each call and why the duty cycle is
therefore low (Fenton 1999).

   LDC bats are very diverse in their echolocation calls, flight behaviour, and
habitat. The calls of most LDC bats are dominated by a broadband FM component
which is well suited to the habitat in which they hunt and their mode of hunting.
Because FM components are good for localization and distinguishing clutter from
target echoes, these bats hunt in habitats where they forage close to vegetation
either in dense vegetation or at the edges of vegetation near the ground or water.
LDC bats can be specialists spending most of their time in dense vegetation
whereas others are more flexible and can move from open habitats to edges of

vegetation, near the ground or water. These bats are able to switch from narrowband calls in open space to broadband calls when hunting close to clutter. Bats with calls dominated by FM components are generally slow manoeuvrable flyers with some flexibility and this is reflected in the shapes of their wings. Some of them have short broad wings for gleaning (taking prey from a substrate) or aerial hawking (catching prey in flight) in dense vegetation while others have intermediate wings for flying in edge habitats (Norberg and Rayner 1987). The FM-dominated calls of some LDC bats that hunt in the open have a quasi-CF component that increases their range of detection. These bats usually have long narrow wings for rapid and agile flight in open habitats, e.g., away from vegetation or high above the ground and water (Schnitzler and Kalko 2001).

2.1.3.2 High-Duty-Cycle (HDC) Bats

High-duty-cycle bats have call durations that are long relative to the silent period between calls (Fenton 1999). These bats are able to have relatively short silent periods and therefore high duration to call period ratios because they avoid self-deafening by separating the emitted call and returning echo in frequency rather than time. High-duty-cycle bats are restricted to the Old World families, Rhinolophidae, Rhinonycteridae, and Hipposideridae and a single species in the Americas, *Pteronotus parnellii*. High-duty-cycle bats emit calls of long duration (up to 100 ms) dominated by a CF component preceded and followed by a brief FM component (Schnitzler and Flieger 1983; Neuweiler 1984). Unlike LDC bats which place most of the call energy in the fundamental, HDC bats place most of the call energy into the second harmonic (Neuweiler 1984; Schnitzler and Kalko 2001). HDC bats have a region in their auditory cortex known as the 'acoustic fovea'. This region has an over-representation of neurons sensitive to a unique and narrow range of frequencies called the reference frequency (Schuller and Pollak 1979). Echoes from the bat's call will return at a slightly higher frequency than the emitted call because of the Doppler effect on frequency as a result of the bats' flight speed relative to that of the target. To ensure that the returning echo falls within the narrow range of frequencies of the acoustic fovea, the bat lowers the frequency of the emitted call a phenomenon known as Doppler-shift compensation (DSC; Fig. 2.1; Schnitzler and Flieger 1983; Neuweiler 1984; Schnitzler and Denzinger 2011). As a result of DSC, the frequency of the emitted call does not therefore fall within the acoustic fovea protecting it from the high intensities at which the calls are emitted and preventing self-deafening. Because the ears are not 'switched off' during echolocation, the bat can emit calls and receive echoes at the same time unlike LDC bats. HDC bats can also hunt from a perch and when stationary emit calls that are 100–300 Hz lower than the reference frequency. These calls are referred to as the 'resting frequency' (RF; Schuller and Pollak 1979).

2.1.4 Tracking Targets

Bats have to adjust their echolocation calls to keep track of a moving object and as the distance between the bat and the object decreases. Bats do so by changing the temporal and spectral parameters of their calls in a systematic way. Echolocation attack sequences can therefore generally be divided into three phases (Griffin et al. 1960): a search phase (narrow bandwidth, long calls with long inter-call intervals), approach phase (shorter, broadband calls and shorter intervals), and the terminal or feeding buzz stage (much shorter calls and intervals with very narrow bandwidth). When a bat searches for prey, it scans as far ahead as possible and as widely as possible to increase the volume of space that it is searching at any one time, maximizing the chances of detecting prey. To do so it uses calls that are of long duration, relatively lower frequency and narrow bandwidth and therefore less affected by atmospheric attenuation. Because the bat is probing further, it takes longer for echoes to return so it emits calls at a slower rate. Longer calls will also increase the probability of at least part of the call being reflected off a distant object as an echo. After detection, the bat switches to the approach phase which allows the bat to classify and localize the object accurately in time and space. This requires greater resolution so it increases the bandwidth of its calls and the call rate while decreasing the duration of the calls. During the approach phase, the bat will have identified the object as edible or something to avoid. If it is prey, the bat homes in on the object switching to the terminal or feeding buzz phase of its attack. During these final stages of the attack, the bat just needs to pinpoint the object in time and space to make an accurate capture. Call rate increases even more, calls get even shorter in duration, and the bandwidth decreases. Calls are now so rapid that instead of discrete calls they appear to be one long buzz of sound, hence the name 'feeding buzz' (Griffin et al. 1960).

2.1.5 Bat Foraging Modes

Bats occupy a diverse range of habitats, and they have evolved different flight modes and echolocation behaviour which allow them to deal with the conditions in these different habitats.

2.1.5.1 Aerial Hawking in Open Habitats

Bats foraging in open areas away from or above vegetation use aerial hawking, i.e. catching flying prey on the wing. Aerial hawkers fly fast to search a large area quickly and to chase down fast-flying prey. They therefore have to probe far in front of them to detect targets at greater distances to give them sufficient time to intercept any target that is detected. In open areas, resolution of targets from each other or from the background vegetation is not necessary and detection and classification are more important. Aerial hawkers use long-duration, narrowband echolocation calls of low frequency, all of which increase detection range. The calls are also narrowly frequency-modulated but aerial hawkers can change both their bandwidth and frequency to improve localization and classification of targets (Neuweiler 1984; Schnitzler and Kalko 2001).

2.1.5.2 Aerial Hawking Near Vegetation

Bats catching insects on the wing near vegetation such as at forest edges and in forest gaps have to distinguish echoes from the target from the much louder echoes coming from the vegetation while keeping track of the vegetation so that they do not collide with it. Bats foraging in such areas use a mixture of very shallow FM or quasi-constant-frequency (QCF) and steep FM calls. Some may also use calls that have both steeply FM and QCF components. Different species of bats hunting in background cluttered space can use different combinations of these types of calls (Schnitzler and Kalko 2001). The narrowband components of these mixed signals allow detection of prey, and the broadband FM components facilitate localization and classification of the background and insect. These bats avoid background masking effects (Fig. 2.2) by searching for and detecting prey in areas where the echo from the prey does not overlap with echoes from the background vegetation. They can also minimize masking effects by using short-duration calls (Neuweiler 1984; Schnitzler and Kalko 2001; Schnitzler et al. 2003).

Dealing with Clutter

Bats hunting in dense vegetation can use aerial hawking, ambush hunting (searching for prey from a perch and intercepting detected prey), and gleaning (taking prey from a substrate, e.g., from a branch or the ground using their mouths or from the surface or just below the surface of water using their feet, fish eagle style) (Fig. 2.3).

2.1.5.3 Gleaning

Bats hunting within dense vegetation (Fig. 2.3) have to be able to resolve target echoes from background clutter and have to deal with both forward (overlap between call and echo) and backward masking (Fig. 2.2) overlap between echoes from the target and background) effects. Target resolution and localization are achieved by using calls of wide bandwidth and high frequencies usually with multiple harmonics. Masking effects are minimized by using calls of short duration, 1–3 ms and maintaining a distance from the target, during localization and classification, that places the target in the overlap-free zone, i.e., where echoes do not

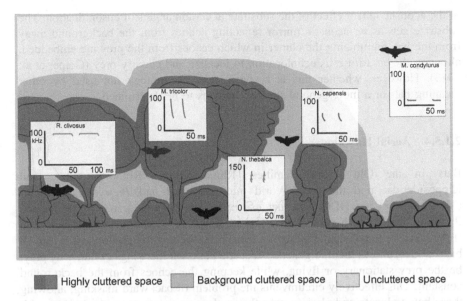

Highly cluttered space   Background cluttered space   Uncluttered space

Fig. 2.3 Bat foraging habitat and duty cycle. Depicted are three habitat categories based on the challenges to echolocation with the wing shape and echolocation calls associated with each category (redrawn from Jacobs 2016). With the exception of *R. clivosus* which is a **high-duty-cycle** echolocator (indicated by the high ratio between call duration and call period —see text) all the other bats are **low-duty-cycle** echolocators (color figure online)

overlap with the emitted call or echoes from the background (Fig. 2.2; Schnitzler and Kalko 2001; Geipel et al. 2013). Because these bats operate in confined spaces, the operation range of their echolocation is short and they are able to use very quiet calls and are often referred to as 'whispering bats' (Schnitzler and Kalko 2001).

Nevertheless, target detection, localization, and classification is still very difficult and almost impossible leading to the suggestion that gleaning bats hunting in clutter can only locate prey by using cues other than echoes such as olfactory (Thies et al. 1998) and visual (Bell 1985; Ekloef and Jones 2003) ones as well as prey-generated sounds such as mating calls (see Chap. 4 and 6; ter Hofstede et al. 2008; Jones et al. 2011) and noises of walking or flying insects (Siemers and Swift 2006; Russo et al. 2007; Jones et al. 2011).

Although it has been demonstrated that some gleaning bats are unable to locate and classify prey in clutter using echolocation alone (Arlettaz et al. 2001) other gleaning species appear to use echolocation throughout a gleaning attack (Schmidt et al. 2000; Swift and Racey 2002; Ratcliffe and Dawson 2003) and some of them appear to be able to localize and classify prey using echolocation alone (Geipel et al. 2013). The gleaning bat *Micronycteris microtis* can do so by combining short-duration, high-frequency broadband calls with three-dimensional hovering flight to minimize the masking effects of echoes from the substrate on which the prey rests (Fig. 2.2). The three-dimensional flight allows the bat to take advantage of the acoustic mirror effect of the substrate at certain angles of ensonification. The substrate acts as an acoustic mirror reflecting echoes from the background away from the bat, minimizing the clutter in which echoes from the prey are embedded, allowing the bat to use its echolocation to localize and classify prey (Geipel et al. 2013). However, whether this is a species-specific strategy not used by other gleaning bats or a more general strategy remains to be determined.

2.1.5.4 Aerial Hawking

Bats in the Old World families, Rhinolophidae, Rhinonycteridae, and Hipposideridae and the New World mustached bat and *Pteronotus parnellii* (Mormoopidae) are HDC bats that use aerial hawking to search for and catch fluttering insects in highly cluttered space close to vegetation or the substrate (Fig. 2.3). The echolocation system of these bats (see sect. 'High-duty-cycle (HDC) bats" above) allows them to generate acoustic glints off the wings of fluttering prey, be the prey stationary or flying, while keeping the echoes from the background constant. This effectively circumvents the problem of backward masking, allowing these bats to locate and classify prey flying close to or sitting on the background. This is achieved by the long-duration CF component of their calls which is therefore clutter insensitive (Schnitzler and Kalko 2001. However, these bats still have to be able to determine the distance to the prey with a degree of precision that allows capture. Such ranging tasks are probably enabled by the FM components at the beginning and end of the CF component (Schnitzler 1968; Schuller et al. 1971; Schnitzler and Denzinger 2011) as is evidenced by increase in the bandwidth and duration particularly of the terminal FM component in response to clutter (Fawcett et al. 2015). Such ranging tasks would also be important for a stationary bat tracking a moving target and explain why even stationary HDC bats include FM components in their calls (Mutumi et al. 2016, S1). These FM components are, however, not clutter insensitive and it is not known how these bats deal with masking effects (Fig. 2.2).

2.1.6 Eavesdropping

Many bats that glean prey use prey-generated acoustic cues, e.g. calls of insects and frogs (Tuttle and Ryan 1981; Ryan et al. 1982; Belwood and Morris 1987; Arlettaz et al. 2001) or rustling noises of insects walking on leaf litter or fluttering their wings (Faure and Barclay 1994; Siemers and Swift 2006; Russo et al. 2007; Jones et al. 2011) for detection, classification, and localization of prey. Many gleaners are characterized by large ears that facilitate passive (i.e. non-echolocating) acoustic localization of prey. However, it appears that gleaners may always emit echolocation signals in flight to determine the position of the site with food, to navigate, and to avoid collisions (e.g. see Ratcliffe et al. 2005) and may use calls of low amplitude to prevent overloading of the hearing system with loud clutter echoes (Schnitzler and Kalko 2001). Passive listening may be a strategy to overcome masking effects that arise from the overlap between target and clutter echoes.

Bats have evolved a plethora of adaptations allowing them to circumvent the acoustic challenges of a diversity of habitats. This has allowed them to prey on an even greater diversity of prey in these habitats, from arthropods to vertebrates. Adaptations to the acoustic challenges of the habitat in which bats forage can confound investigations into the co-evolution between bats and their prey. This is also true of prey defences. The identification of co-evolved traits is therefore dependent on the extent to which we can separate the influence of habitat from that of the predator-prey interaction on the evolution of bats and their insect prey.

References

Arlettaz R, Jones G, Racey PA (2001) Effect of acoustic clutter on prey detection by bats. Nature 414(6865):742–745

Bates ME, Simmons JA, Zorikov TV (2011) Bats use echo harmonic structure to distinguish their targets from background clutter. Science 333(6042):627–630

Bell GP (1985) The sensory basis of prey location by the California leaf-nosed bat *Macrotus californicus* (Chiroptera: Phyllostomidae). Behav Ecol Sociobiol 16(4):343–347

Belwood JJ, Morris GK (1987) Bat predation and its influence on calling behavior in neotropical katydids. Science 238(4823):64–67

Boonman A, Bumrungsri S, Yovel Y (2014) Nonecholocating fruit bats produce biosonar clicks with their wings. Curr Biol 24(24):2962–2967

Chiu C, Moss CF (2007) The role of the external ear in vertical sound localization in the free flying bat, *Eptesicus fuscus*. J Acoust Soc Am 121(4):2227–2235

Ekloef J, Jones G (2003) Use of vision in prey detection by brown long-eared bats, *Plecotus auritus*. Anim Behav 66:949–953

Faure PA, Barclay RMR (1994) Substrate-gleaning versus aerial-hawking: plasticity in the foraging and echolocation behaviour of the long-eared bat, myotis evotis. J Comp Physiol A Neuroethology Sens Neural Behav Physiol 174(5):651–660

Fawcett K, Jacobs DS, Surlykke A, Ratcliffe JM (2015) Echolocation in the bat, *Rhinolophus capensis*: the influence of clutter, conspecifics and prey on call design and intensity. Biol Open 4(6):693–701

Fenton MB (1999) Describing the echolocation calls and behaviour of bats. Acta Chiropterologica 1(2):127–136

Fenton MB, Ratcliffe JM (2014) Sensory biology: echolocation from click to call, mouth to wing. Curr Biol 24(24):4

Fenton MB, Jacobs DS, Richardson EJ, Taylor PJ, White E (2004) Individual signatures in the frequency-modulated sweep calls of African large-eared, free-tailed bats *Otomops martiensseni* (Chiroptera: Molossidae). J Zool 262:11–19

Fullard JH (1987) Sensory ecology and neuroethology of moths and bats: interaction in a global perspective. Cambridge University Press, Cambridge

Funakoshi K, Zubaid A, Matsumura S (1995) Regular pulse emission in some megachiropteran bats. Zoolog Sci 12(4):503–505

Geipel I, Jung K, Kalko EKV (2013) Perception of silent and motionless prey on vegetation by echolocation in the gleaning bat *Micronycteris microtis*. Proc Roy Soc Lond B Biol Sci 280 (1754):7

Goerlitz HR, ter Hofstede HM, Zeale MRK, Jones G, Holderied MW (2010) An aerial-hawking bat uses stealth echolocation to counter moth hearing. Curr Biol 20(17):1568–1572

Goudy-Trainor A, Freeman PW (2002) Call parameters and facial features in bats: a surprising failure of form following function. Acta Chiropterologica 4(1):1–16

Gould E (1988) Wing-clapping sounds of *Eonycteris spelaea* (Pteropodidae) in Malaysia. J Mammal 69(2):378–379

Griffin DR (1944) Echolocation by blind men, bats and radar. Science 100(2609):589–590. doi:10.1126/science.100.2609.589

Griffin DR (1958) Listening in the dark: the acoustic orientation of bats and men. Yale University Press, New Haven

Griffin DR, Webster FA, Michael CR (1960) The echolocation of flying insects by bats. Anim Behav 8(3):141–154

Halfwerk W, Dixon MM, Ottens KJ, Taylor RC, Ryan MJ, Page RA, Jones PL (2014) Risks of multimodal signaling: bat predators attend to dynamic motion in frog sexual displays. J Exp Biol 217(17):3038–3044

Holland RA, Waters DA, Rayner JMV (2004) Echolocation signal structure in the megachiropteran bat *Rousettus aegyptiacus* Geoffroy 1810. J Exp Biol 207(25):4361–4369

Jacobs DS (2000) Community level support for the allotonic frequency hypothesis. Acta Chiropterologica 2(2):197–207

Jacobs DS (2016) Evolution's chimera: bats and the marvel of evolutionary adaptation. University of Cape Town Press, Cape Town

Jones PL, Page RA, Hartbauer M, Siemers BM (2011) Behavioral evidence for eavesdropping on prey song in two palearctic sibling bat species. Behav Ecol Sociobiol 65(2):333–340

Lawrence BD, Simmons JA (1982) Measurements of atmospheric attenuation at ultrasonic frequencies and the significance for echolocation by bats. J Acoust Soc Am 71(3):585–590

Luo J, Koselj K, Zsebok S, Siemers BM, Goerlitz HR (2013) Global warming alters sound transmission: differential impact on the prey detection ability of echolocating bats. J Roy Soc Interface/Roy Soc 11:20130961. doi:10.1098/rsif.2013.0961 PMID: 24335559

Matthias K, Herkta B, Barnikela G, Skidmoreb AK, Fahr J (2016) A high-resolution model of bat diversity and endemism for continental Africa. Ecol Model 320:9–28

Moss CF, Schnitzler H-U (1995) Behavioral studies of auditory information processing. In: Popper AN, Fay RR (eds) Hearing by bats, vol 5. Springer, New York, pp 87–145

Mutumi GL, Jacobs DS, Winker H (2016) Sensory drive mediated by climatic gradients partially explains divergence in acoustic signals in two horseshoe bat species, *Rhinolophus swinnyi* and *Rhinolophus simulator*. PLoS ONE 11(1):e0148053. doi:10.1371/journal.pone.0148053

Neuweiler G (1984) Foraging, echolocation and audition in bats. Naturwissenschaften 71(9):446–455

Neuweiler G (1989) Foraging ecology and audition in echolocating bats. Trends Ecol Evol 4(6):160–166

Neuweiler G (1990) Auditory adaptations for prey capture in echolocating bats. Physiol Rev 70(3):615–641

Norberg UM, Rayner JMV (1987) Ecological morphology and flight in bats (Mammalia; Chiroptera): wing adaptations, flight performance, foraging strategy and echolocation. Philos Trans Roy Soc Lond B 316:335–427

Novick A (1958) Orientation in paleotropical bats. II. Megachiroptera. J Exp Zool 137(3):443–461

Obrist MK, Fenton MB, Eger JL, Schlegel PA (1993) What ears do for bats - a comparative-study of pinna sound pressure transformation in Chiroptera. J Exp Biol 180:119–152

Payne RS (1971) Acoustic location of prey by barn owls (*Tyto alba*). J Exp Biol 54:535–573

Pedersen SC (1993) Cephalometric correlates of echolocation in the Chiroptera. J Morphol 218(1):85–98

Pye JD (1993) Is fidelity futile? The 'true' signal is illusory, especially with ultrasound. Bioacoustics 4(4):271–286

Raghuram H, Gopukumar N, Sripathi K (2007) Presence of single as well as double clicks in the echolocation signals of a fruit bat, *Rousettus leschenaulti* (Chiroptera: Pteropodidae). Folia Zool 56(1):33–38

Ratcliffe JM, Dawson JW (2003) Behavioural flexibility: the little brown bat, *Myotis lucifugus*, and the northern long-eared bat, *M. septentrionalis*, both glean and hawk prey. Anim Behav 66:847–856

Ratcliffe JM, Raghuram H, Marimuthu G, Fullard JH, Fenton MB (2005) Hunting in unfamiliar space: echolocation in the Indian false vampire bat, *Megaderma lyra*, when gleaning prey. Behav Ecol Sociobiol 58:157–164

Rhebergen F, Taylor RC, Ryan MJ, Page RA, Halfwerk W (2015) Multimodal cues improve prey localization under complex environmental conditions. Proc Roy Soc B Biol Sci 282(1814)

Russo D, Jones G, Arlettaz R (2007) Echolocation and passive listening by foraging mouse-eared bats *Myotis myotis* and *M blythii*. J Exp Biol 210(1):166–176

Ryan MJ, Tuttle MD, Rand AS (1982) Bat predation and sexual advertisement in a Neotropical Anuran. Am Nat 119(1):136–139

Schmidt S (1988) Discrimination of target surface structure in the echolocating bat, *Megaderma lyra* animal sonar. Springer, pp 507–511

Schmidt S, Hanke S, Pillat J (2000) The role of echolocation in the hunting of terrestrial prey—new evidence for an underestimated strategy in the gleaning bat, *Megaderma lyra*. J Comp Physiol A Neuroethology Sens Neural Behav Physiol 186(10):975–988

Schnitzler H-U (1968) Die Ultraschall-Ortungslaute der Hufeisen fledermaeuse (Chiroptera – Rhinolophidae) in verschiedenen Orientierungssituationen. Z. Vgl. Physiol. 57:376–408

Schnitzler H-U, Denzinger A (2011) Auditory fovea and Doppler shift compensation: adaptations for flutter detection in echolocating bats using CF-FM signals. J Comp Physiol A Neuroethology Sens Neural Behav Physiol 197(5):541–559

Schnitzler H-U, Flieger E (1983) Detection of oscillating target movements by echolocation in the greater horseshoe bat. J Comp Physiol 153(3):385–391

Schnitzler H-U, Kalko EKV (2001) Echolocation by insect-eating bats. Bioscience 51(7):557–569

Schnitzler H-U, Moss CF, Denzinger A (2003) From spatial orientation to food acquisition in echolocation bats. Trends Ecol Evol 18(8):386–394

Schoeman MC, Goodman SM (2012) Vocalizations in the Malagasy cave-dwelling fruit bat, *Eidolon dupreanum*: possible evidence of incipient echolocation? Acta Chiropterologica 14 (2):409–416

Schuller G, Pollak G (1979) Disproportionate frequency representation in the inferior colliculus of Doppler-compensating greater horseshoe bats—evidence for an acoustic fovea. J Comp Physiol 132(1):47–54

Schuller G, Neuweiler G, Schnitzler H-U (1971) Collicular responses to the frequency modulated final part of echolocation sounds in Rhinolophus ferrrumequinum. Zeitschrift fuer vergleichende Physiologie 74:153–155

Siemers BM, Swift SM (2006) Differences in sensory ecology contribute to resource partitioning in the bats *Myotis bechsteinii* and *Myotis nattereri* (Chiroptera: Vespertilionidae). Behav Ecol Sociobiol 59(3):373–380

Simon R, Knörnschild M, Tschapka M, Schneider A, Passauer N, Kalko EKV, von Helversen O (2014) Biosonar resolving power: echo-acoustic perception of surface structures in the submillimeter range. Front Physiol 5

Surlykke A, Kalko EKV (2008) Echolocating bats cry out loud to detect their prey. PLoS ONE 3 (4):10

Swift SM, Racey PA (2002) Gleaning as a foraging strategy in Natterer's bat *Myotis nattereri*. Behav Ecol Sociobiol 52(5):408–416

ter Hofstede HM, Ratcliffe JM, Fullard JH (2008) The effectiveness of katydid (*Neoconocephalus ensiger*) song cessation as antipredator defence against the gleaning bat *Myotis septentrionalis*. Behav Ecol Sociobiol 63(2):217–226

Thies W, Kalko EKV, Schnitzler H-U (1998) The roles of echolocation and olfaction in two neotropical fruit-eating bats, *Carollia perspicillata* and *C. castanea*, feeding on piper. Behav Ecol Sociobiol 42(6):397–409

Tuttle MD, Ryan MJ (1981) Bat predation and the evolution of frog vocalizations in the Neotropics. Science 214(4521):677–678

von Der Emde G, Menne D (1989) Discrimination of insect wingbeat-frequencies by the bat *Rhinolophus ferrumequinum*. J Comp Physiol A Neuroethology Sens Neural Behav Physiol 164:663–671

von Der Emde G, Schnitzler H-U (1990) Classification of insects by echolocating greater horseshoe bats. J Comp Physiol A Neuroethology Sens Neural Behav Physiol 167(3):423–430

Young AT (1981) Rayleigh scattering. Appl Opt 20(4):533–535

Yovel Y, Falk B, Moss CF, Ulanovsky N (2011) Active control of acoustic field-of-view in a biosonar system. PLoS Biol 9(9):10

Zhuang Q, Müller R (2006) Noseleaf furrows in a horseshoe bat act as resonance cavities shaping the biosonar beam. Phys Rev Lett 97(21):218701

Chapter 3
Non-auditory Defences of Prey Against Bat Predation

Abstract Insects have evolved a variety of traits that appear to be direct countermeasures to bat echolocation. Insect defences can be grouped into primary defences to avoid detection and secondary defences that allow them to escape being captured after the bat has detected them. These defences include auditory and non-auditory defences as wells as ultrasound in the form of tymbal clicks. Here, we review the non-auditory adaptations of insects, including anti-bat traits such as body size, morphological or acoustic crypsis, interference with the echolocation signal by some body appendages and scales in moths, the detection of wind caused by bat wing beats, and increased detection capabilities through large eyes. Mechanisms for avoidance of detection by bats include shifting activity phases, temporally shifting their habitat in response to bat activity, flightlessness or swamping the predator, and confusing its detection system by occurring in aggregations, for example on leks. We discuss bat predation and other factors that affect lek size. Finally, we also consider whether these apparent anti-bat traits could have evolved for reasons other than bat predation, for example flightlessness in moths could have evolved for increased fecundity.

3.1 Introduction

3.1.1 Primary Versus Secondary Defences

Prey defences can be placed into two broad categories. Primary defences are defences which allow the prey to avoid detection by predators and usually operate before the predator is aware of the prey's presence. Secondary defences are defences that operate once the predator has detected the prey and the prey is aware of the predator (Staudinger et al. 2011; Conner 2014). The function of secondary

D.S. Jacobs and A. Bastian, *Predator–Prey Interactions: Co-evolution between Bats and their Prey*, SpringerBriefs in Animal Sciences,
DOI 10.1007/978-3-319-32492-0_3

defences is to improve the chances of the prey surviving an encounter with the predator (Robinson 1969). In bat–prey interactions, secondary defences are mainly mediated by audition because the main means that bats have to detect prey is their echolocation and prey exploit this. Primary defences, in contrast, are usually restricted to prey that do not have ears. Although bats and insects have been engaged in acoustic interactions over more than 60 million years, close to half of moth species do not have bat-detecting ears (Soutar and Fullard 2004; Barber et al. 2015) and are vulnerable to bat predation. Amongst the Coleoptera audition is thus far only known in the families Scarabaeidae and Cicindelidae (Spangler 1988; Forrest et al. 1997). Other nocturnal insects such as the Diptera, Trichoptera, Ephemeroptera, Plecoptera, Isoptera, Hymenoptera, and Heteroptera are not known to have ears. However, insects that lack ears have a variety of defences against bat predation that include both morphological and behavioural adaptations.

3.2 Physical Adaptations

3.2.1 Size

Insects with wing lengths of <3 mm are seldom found in bat diets. This may be because such small insects may be too small to be detected by bat echolocation. Alternatively, such small insects may be detectable by bats but are ignored because they may be energetically unprofitable for bats to prey upon (Jacobs 1999; Houston et al. 2004). Similarly, large insects, e.g. earless moths in the families Sphingidae and Saturniidae, which can reach the size of some bats, may be too big to be handled by bats that catch prey on the wing and may therefore be ignored despite being very detectable. However, large size may also make these large earless Macrolepidoptera difficult to capture. They have significantly larger wingspans and wing loadings than eared forms allowing them to fly fast and erratically making it difficult for bats to capture them. This is supported by the fact that these earless moths also have higher metabolic rates than eared Macrolepidoptera. The higher metabolic rate may be an adaptation allowing fast erratic flight and part of an evolved defence against bat predation that does not involve hearing (Rydell and Lancaster 2000). However, size, metabolic rate and their relationship with fast and erratic flight should be investigated in the context of bat predation in eared hawkmoths (Sphingidae; Roeder et al. 1968; Göpfert et al. 2002). Such an investigation should correct for body size and phylogeny and should consider potential developmental trade-offs, e.g. energy used to develop ears may not be available for fast flight and vice versa (Michael Greenfield, pers. comm.). Although large, hawkmoths have ears suggesting that large size does not always confer immunity to bat attacks (Jones and Rydell 2003; see also acoustic deflection below).

3.2.2 Cryptic Shapes/Texture/Material

Most examples of crypsis, based on body shape, as a defence against predation in the animal world involves visually orientated predators and the possibility that the shape of the body of prey may be acoustically cryptic has received little attention. Conner (2014) provides an in depth review of acoustic crypsis and other acoustic anti-predator strategies in insects. It is reasonable to expect that if the shape of prey resembles a non-prey item, for example a twig (stick insects, Phasmatodea) or a leaf (Walking leaf insects, Phylliidae), that echoes from such prey would also resemble a twig or leaf, respectively, to most bats. However, presumably a grasshopper camouflaged as a leaf (mostly by colour rather than by shape) to a visually oriented predator might nevertheless still produce an echo that resembles a grasshopper to a gleaning bat, provided the bat can deal with the masking effects of pulse-echo overlap (Fig. 2.2) and the echoes from the background on which the grasshopper is perched (see Geipel et al. 2013). In contrast, none of these defences are likely to work on bats that are able to use shape, surface structure, and material to identify immobile prey (e.g. *Micronycteris microtis*, Geipel et al. 2013). Such bats are able to detect and eat stick insects albeit in low proportions (Kalka and Kalko 2006).

3.2.3 Body Appendages for Acoustic Diversion

Adaptations to divert the attacks of visually guided predators from essential body parts to less essential parts have evolved repeatedly in animals, e.g. brightly coloured tails in lizards and eyespots in moths and butterflies. It has recently been shown that these colours and eyespots deflect the attack of visually oriented predators away from essential body parts allowing the organism to escape and survive a predatory attack (Olofsson et al. 2010; Watson et al. 2012). An acoustic version of an attack-deflecting morphological character has recently been discovered in luna moths (*Actias luna*; Saturniidae; Barber et al. 2015). The hind wings of these moths end in long spatulate tails which apparently exert little influence on the flight dynamics of these moths. However, when these moths fly, these tails spin and generate an acoustic diversion attracting the attacks of echolocating bats to the less essential tails and away from the more essential parts of the moth's body. The trajectory of a bat attack on intact luna moths versus moths with their tails removed was different. The bats attacked the tails of the intact luna moths rather than their bodies, but attacked the bodies of moths that had their tails removed. Bats captured 47% more of the moths with their tails removed than they did intact moths which means that the tails provided similar levels of protection as hearing does in eared moths (see Chap. 4 and 5). Although difference in size between intact and tailless moths did influence differences in the survival rate of the two groups of moth, difference in size could not explain all of the variation in survival rates. Exactly how the spinning moth tails deflect the attack from essential body parts is not known.

Ensonification of the spinning tails indicate that they produce distinct wing-like amplitude and frequency modulations on the returning echo from frequency-modulated (FM) signals but whether the bats detect these as alternative targets or more conspicuous elements of a single target is unknown (Barber et al. 2015).

3.2.4 Large Eyes

Some earless moths (Sphingidae) have relatively larger eyes than their eared relatives, and it has been proposed that they may use vision to detect bats (Soutar and Fullard 2004).

Sphingids frequently consume nectar (O'Brien 1999) which they obtain from flowers while hovering (Wicklein and Strausfeld 2000). Some eared sphingids supposedly use their hearing while hovering to detect attacking bats (Roeder et al. 1968), so it is possible that earless sphingids use their relatively larger eyes to see bats. However, there is very little direct evidence that these moths use their eyes to alert them to the presence of an attacking bat or that their eyes have increased in direct response to bat predation.

3.2.5 Cerci: Detection of Wind from the Wings of Bats

Cerci are paired appendages on the rear-most segments of many arthropods, including insects, and many of them have a sensory function. They contain many hair types and are sensitive to mechanical stimulation caused by air movement (Boyan and Ball 1990). They are known to be implicated in escape responses from terrestrial predators in cockroaches (Camhi 1984; Ritzmann 1984) and crickets (Tauber and Camhi 1995) and may also be implicated in escape responses from aerial predators (Ganihar et al. 1994; Triblehorn et al. 2008; Triblehorn and Yager 2006). Evasive behaviours mediated by the detection of wind generated by the flapping wings of an approaching bat (or bird) by the cerci of insects may contribute to successful escape from predation. Given that such wind stimuli would only be effective over short distances whether the cerci increase the escape rate of insects would depend on whether the insect is given sufficient warning to execute evasive manoeuvres. Evidence for a defensive function for cerci against bat predation is equivocal. Deafened mantids with functioning cerci did not escape more often than deafened mantids with deactivated cerci, and there was no statistical difference in the frequency with which bats dropped mantids with functioning cerci and mantids with deactivated cerci. However, dropped mantids always survived the encounter, and the latter result may have been influenced by small sample sizes (Triblehorn et al. 2008). In another experiment, Triblehorn and Yager (2006) trained the aerial hawking bat, *Eptesicus fuscus*, to prey on tethered mantids (Parasphendale agrionina, Mantidae) while the nerve impulses in the abdomen were recorded.

The recordings showed that the mantid could detect the bat 74 ms before the bat made contact with the mantid. This was estimated to provide the mantid with 36 ms to perform evasive manoeuvres to avoid capture. Although this is sufficient time for a response, it probably may not be enough time for the mantid to completely avoid the bat but may be enough time for a response that would make the bat mishandle the mantid causing it to drop the mantid.

In a similar experiment on tettigoniids, Hartbauer et al. (2010), using extracellular recordings, showed that the dorsal giant interneurons of these katydids fired regular bursts in response to pulses of ultrasound designed to mimic an attacking bat. More importantly, the time elapsed between the first response of wind-sensitive interneurons and potential contact with a predatory bat was 860 ms much longer than that found in mantids. This time interval corresponds to a distance of ca. 72 cm and suggests that wind-sensitive cerci may enable earless insects to detect bats or allow eared insects to execute evasive manoeuvres to escape bats that went undetected either because the insect auditory system was circumvented by eavesdropping bats or by the insect's own singing to attract mates (see Chap. 4).

3.3 Behavioural Adaptations

3.3.1 Timing of Diel Activity

Bats are active only between dusk and dawn and this predictable activity may be exploited by prey. Many arthropods avoid bat predation by being partly or entirely diurnal (e.g. butterflies, Papilionoidea, and Hesperioidea; Fullard and Napoleone 2001) or crepuscular e.g. lekking moths that lek for short periods at dusk (Andersson et al. 1998). Seemingly maladaptive, many earless moths are almost entirely nocturnal when bats are the most active. However, they tend to fly close to the substrate where backward masking effects of background clutter (the echo from the background mask the echo from the moths) make them less conspicuous to aerial hunting bats and/or they are bigger and use fast erratic flight allowing them to avoid detection and capture by most bats (Rydell and Lancaster 2000; Fullard and Napoleone 2001).

Some earless moth species are also more active in parts of summer when bat activity has not reached its peak. In a study in Canada on moth activity, most Saturniidae and Lasiocampidae moths were most active in the early part of the northern summer (May and June) before bat activity reached its peak in July and August. It thus seems that some families of deaf moths avoid bat predation by temporal partitioning of the habitat. In contrast, the activity of some hawkmoths matched those of bats (Yack 1988). The coincidence of hawkmoth and bat activity might be explained by some hawkmoths not being deaf. Their hearing may provide adequate defence against bats allowing them to be active at the same time as their predators.

Despite observations that earless insects have diel activity patterns that allow them to avoid bats, it is by no means certain that these activity patterns evolved in direct response to bat predation. For example, the nocturnal flight activity of earless butterflies (Nymphalidae) in a bat-free habitat on the Pacific island of Moorea did not differ significantly from the nocturnal activity of nymphalids in the bat-inhabited habitat of Queensland, Australia. Thus, release from bat predation on Moorea did not produce significantly greater nocturnal activity neither did exposure to bat predation in Queensland lead to greater diurnal activity (Fullard 2000). There may be several explanations for this, e.g. bat predation does not influence diel activity of their insect prey, or if it does, there is still sufficient gene flow between bat-exposed and bat-free habitats to avoid shifts in diel activity. Alternatively, physiological constraints as a result of adaptations to one or the other temporal habitat may prevent such shifts. These alternatives require rigorous investigation.

3.4 Acoustic Concealment (Crypsis)

Although previously defined in the context of visually oriented predators, the definition of crypsis has been extended beyond visual traits to include all traits (visual, chemical, tactile, electric, and acoustic cues) that minimize the probability of an organism being detected when potentially detectable by an observer (Conner 2014).

3.4.1 Timing of Flight, Flight Patterns

Earless moths, without early warning sensory defences such as hearing, may be limited to passive anti-bat flight adaptations such as reduced flight. There is some evidence that earless moths fly less during the night than eared moths (Morrill and Fullard 1992; Soutar and Fullard 2004). Reduced flight means that the moths are motionless in vegetation for longer periods of time than eared moths and are therefore less detectable to aerial hawking and gleaning bats that have echolocation systems that are not clutter tolerant (Fenton 1990; Schnitzler and Kalko 2001). Being motionless and silent also protects them from gleaning bats that use prey generated sounds to locate prey (Arlettaz et al. 2001).

3.4.2 Acoustic Concealment

Members of the Hepialidae family of moths do not have ultrasonic hearing but males are conspicuous during their display flight on leks and bats (e.g. *Eptesicus nilssonii*) prey upon them. One species of hepialids, *Hepialus humuli*, uses at least two anti-bat defences. Firstly, they avoided most bats by restricting their display flight to only brief periods at dusk. However, they were still exposed to aerial hawking bats that emerged early. To avoid these bats, they employed acoustic crypsis by flying close enough to the vegetation that any echoes generated off their bodies by bat echolocation are concealed in background echoes from the vegetation. Moths that sometimes left the safety of the vegetation were heavily preyed upon by bats (Rydell 1998).

Some moths also use very low intensity courtship songs to avoid detection by eavesdropping predators. The Asian corn borer moth, *Ostrinia furnacalis*, scrape specialized scales on its wing against those on the thorax to produce a courtship song at very low intensity (46 dB SPL at 1 cm). The frequencies in these songs range from 40–80 kHz, the same range of the echolocation calls of the most common sympatric bats. The low intensities are therefore necessary to avoid detection by bats. The female moth is able to hear the male's song because he sings directly into her ear. Because the song is at the same frequencies used by echolocating bats, it initiates immobility in the female, a strategy used to avoid detection by bats. The female's immobility allows the male to mate successfully with her. This is an example of mimicry of a predator to increase reproductive success. The low intensity of the male's song avoids discovery by eavesdropping competitive conspecific males (Nakano et al. 2009) contributing to the singing male's mating success and also protects the couple from detection by eavesdropping bats (Nakano et al. 2008).

Some insects may decrease the amplitude of the echo they generate from an impinging bat call preventing the bat from detecting them. The scales covering the body of some moths may function as sound absorbers reducing the amount of the sound reflected off their bodies as echoes (Moss and Zagaeski 1994; Roeder 1998), and at least one study has shown that at some frequencies of sound, the moth body absorbs more sound than the bodies of beetles (Moss and Zagaeski 1994). This effect was more pronounced when comparing moth wings with scales, moth wings without scales, and butterfly wings (Zeng et al. 2011). The scales of the nocturnal moth *Spilosoma niveus* and *Rhyparoides amurensis* (Erebidae) ensonified with sounds between 40 and 60 kHz, frequencies used by their most common sympatric bats, more than doubled the absorption of sound by the wings compared to moth wings without scales and the wings of insects (butterflies) not exposed to bat predation. Moth scales decreased the echo intensity by up to 2 dB over wings without scales reducing the detection distance at which a bat detects the moth by 5–6%. This gives the moth a small but significant advantage in avoiding detection (Zeng et al. 2011). Moth scales appear to have properties similar to sound-absorbing materials. They are composed of honeycomb-like hollows between them and are typically covered with micropores and trabeculae, properties not present in butterfly scales (Zeng et al. 2011).

Similarly, other insects e.g. field crickets demonstrate a variety of primary, or preventative, anti-predator defences, such as moving under dense vegetation (Hedrick and Dill 1993), calling from burrows (Bailey and Haythornthwaite 1998), and modifying their behaviour depending on the predation or parasitoid risk of a particular habitat (Hedrick and Kortet 2006; Zuk et al. 2006; Bailey et al. 2008). Some of these behaviours may also serve other functions, for example, mole crickets may call from burrows because the burrow acts as an acoustic horn amplifying their calls for detection by potential mates (Bennet-Clark 1987; Prestwich 1994; Hill et al. 2006) but nevertheless also protecting the insect from predation.

3.5 Habitat Shifts

The water strider (*Aquarius najas*; Hemiptera: Gerridae) is a bug that does not employ anti-bat defences such as ultrasonic hearing or unpalatability. However, these bugs were seldom taken by bats (*Myotis daubentonii*) even though these bugs dominated the insect fauna in streams over which the bats foraged in southern Sweden. These bugs used acoustic concealment by shifting their habitat temporally in response to bat activity. When bats were active, the bugs stayed motionless within 1 m of the stream bank possibly hiding in the acoustic clutter of the bank. When bats were absent, they moved into the centre of the stream (Svensson et al. 2002).

3.6 Flightlessness

Some insects avoid aerial hunting bats by not flying at all. Bats may be responsible for the evolution of flightless varieties or genders in lineages that consist mostly of volant organisms. The females of some moth species e.g. winter moths (Geometridae) and Gypsy moths (Lymantriidae) are functionally deaf, have reduced wings and are flightless (Rydell et al. 1997; Cardone and Fullard 1988). The males of these moths fly and have sophisticated ears that are used to detect the echolocation calls of sympatric bats. Flightlessness in females, with subsequent degeneration of the ears, may have evolved as a defence against bat predation, at least from aerial hawking bats, because such females are concealed by vegetation. However, bat predation may not be the sole selection pressure producing flightlessness because reduction of the ears leaves more space in the females' abdomen for the production of eggs. Since female size (and therefore the abdominal space available for the production of eggs) in geometrid moths is positively correlated with fecundity (Haukioja and Neuvonen 1985), flightlessness in females may also have evolved because it increases egg production (Rydell et al. 1997).

Flightlessness has also evolved in arthropods, other than moths, that are also preyed upon by bats e.g. in at least one gender in crickets (usually the males) (Pollack and Martins 2007) and some mantises (usually females) (Yager 1990). In the Pacific field cricket, *Teleogryllus oceanicus*, (Orthoptera: Gryllidae) it is the females that fly and the males that are flightless. Males attract females by singing from the ground where they are protected from bat predation by the echo clutter produced by the vegetation. The females have functional ears tuned to the echolocation frequencies of sympatric bats and show negative phonotaxis, mediated by the auditory neuron (AN2), when exposed to the calls of sympatric aerial-hawking bats (Fullard et al. 2005). However, when grounded males and females were tested with the calls of a gleaning bat, *Nyctophilus geoffroyi*, the males did not stop singing neither did the females stop crawling even though the bat echolocation calls elicited action potentials at high firing rates in the AN2 of the crickets. Thus, although the crickets could hear the bat calls, they did not respond behaviourally to the threat of predation. It is possible that a behavioural response on the ground to predation from a gleaning bat may not have evolved because on the ground these crickets are protected by primary behavioural defences (calling from burrows or vegetation). Alternatively, it could be related to a switch in function of the AN2 from predator detection in the air to locating conspecifics on the ground, in combination with the AN1 auditory interneuron (ter Hofstede et al. 2009).

Although flightlessness does confer protection on many arthropods that are normally taken by bats, there is evidence of other evolutionary responses to bat predation in these organisms (e.g. negative phonotaxis in the aerial form). However, it is not known for most of these organisms whether bat predation was the primary selection pressure. However, flightlessness is usually associated with hearing degeneration and is therefore not associated with e.g. courtship which would not predict an association between hearing loss and flightlessness.

3.7 Predator Swamping

There is a limit to the number of prey a predator can catch and handle per unit of time, and a sudden encounter with a large number of prey, such as on a lek, may swamp or satiate the predator population, thereby reducing the fraction of prey taken by predators (Pulliam and Caraco 1984; Ims 1990). Such predator swamping may dilute the risk of predation faced by any individual on the lek (Turchin and Kareiva 1989). Males of the lekking moth, *Achroia grisella*, benefit from being part of larger leks through these being more attractive to females and decreasing the probability of being the one taken by bats. Although leks are preyed upon by bats (*Rhinolophus ferrumequinum*), the incidences of predation do not increase in larger leks (Alem et al. 2011). Furthermore, lek size in *A. grisella* is limited by other factors such as preference by males for smaller leks and limits to female cognition. The latter may have selected against male preference for leks that had 4 or more males. Females distinguish among leks by their overall song rate and females appear to be neuro-ethologically incapable of distinguishing among the higher call rates produced by leks comprising four or more males (Alem et al. 2015). Thus, lek sizes appear to be determined by the neuroethology of moths rather than by bat predation but moths on leks nevertheless benefit from the dilution effect.

References

Alem S, Koselj K, Siemers BM, Greenfield MD (2011) Bat predation and the evolution of leks in acoustic moths. Behav Ecol Sociobiol 65(11):2105–2116

Alem S, Clanet C, Party V, Dixsaut A, Greenfield MD (2015) What determines lek size? cognitive constraints and per capita attraction of females limit male aggregation in an acoustic moth. Anim Behav 100:10–115

Andersson S, Rydell J, Svensson MGE (1998) Light, predation and the lekking behaviour of the ghost swift *Hepialus humuli* (L.) (Lepidoptera: Hepialidae). Proc Roy Soc Lond B Biol Sci 265 (1403):1345–1351

Arlettaz R, Jones G, Racey PA (2001) Effect of acoustic clutter on prey detection by bats. Nature 414(6865):742–745

Bailey WJ, Haythornthwaite S (1998) Risks of calling by the field cricket *Teleogryllus oceanicus*; potential predation by Australian long-eared bats. J Zool 244(04):505–513

Bailey NW, McNabb JR, Zuk M (2008) Pre-existing behavior facilitated the loss of a sexual signal in the field cricket *Teleogryllus oceanicus*. Behav Ecol 19(1):202–207

Barber JR, Leavell BC, Keener AL, Breinholt JW, Chadwell BA, McClure CJW, Hill GM, Kawahara AY (2015) Moth tails divert bat attack: evolution of acoustic deflection. Proc Natl Acad Sci USA 112(9):2812–2816

Bennet-Clark HC (1987) The tuned singing burrow of mole crickets. J Exp Biol 128:383–409

Boyan GS, Ball EE (1990) Neuronal organization and information processing in the wind-sensitive cercal receptor/giant interneurone system of the locust and other orthopteroid insects. Prog Neurobiol 35(3):217–243

Camhi JM (1984) Neuroethology: nerve cells and the natural behavior of animals, 1st edn. Sinauer Associates Sunderland, Massachusetts

Cardone B, Fullard JH (1988) Auditory characteristics and sexual dimorphism in the gypsy moth. Physiol Entomol 13(1):9–14

Conner WE (2014) Adaptive sounds and silences: acoustic anti-predator strategies in insects. In: Hedwig B (ed) Insect hearing and acoustic communication, vol 1. Animal signals and communication, Springer, Berlin, pp 65–79

Fenton MB (1990) The foraging behavior and ecology of animal-eating bats. Can J Zool Rev Can Zool 68(3):411–422

Forrest TG, Read MP, Farris HE, Hoy RR (1997) A tympanal hearing organ in scarab beetles. J Exp Biol 200:601–606

Fullard JH (2000) Day-flying butterflies remain day-flying in a polynesian, bat-free habitat. Proc Roy Soc Lond B Biol Sci 267(1459):2295–2300

Fullard JH, Napoleone N (2001) Diel flight periodicity and the evolution of auditory defences in the Macrolepidoptera. Anim Behav 62:349–368

Fullard JH, Ratcliffe JM, Guignion C (2005) Sensory ecology of predator-prey interactions: responses of the AN2 interneuron in the field cricket, *Teleogryllus oceanicus* to the echolocation calls of sympatric bats. J Comp Physiol A Neuroethology Sens Neural Behav Physiol 191(7):605–618

Ganihar D, Libersat F, Wendler G, Camhi JM (1994) Wind-evoked evasive responses in flying cockroaches. J Comp Physiol A Neuroethology Sens Neural Behav Physiol 175(1):49–65

Geipel I, Jung K, Kalko EKV (2013) Perception of silent and motionless prey on vegetation by echolocation in the gleaning bat *Micronycteris microtis*. Proc Roy Soc Lond B Biol Sci 280 (1754):7

Göpfert MC, Surlykke A, Wasserthal LT (2002) Tympanal and atympanal 'mouth-ears' in hawkmoths (Sphingidae). Proc Roy Soc Lond B Biol Sci 269(1486):89–95

Hartbauer M, Ofner E, Grossauer V, Siemers BM (2010) The cercal organ may provide singing tettigoniids a backup sensory system for the detection of eavesdropping bats. PLoS ONE 5 (9):13

Haukioja E, Neuvonen S (1985) The relationship between size and reproductive potential in male and female *Epirrita autumnata* (Lep, Geometridae). Ecol Entomol 10(3):267–270

Hedrick AV, Dill LM (1993) Mate choice by female crickets is influenced by predation risk. Anim Behav 46(1):193–196

Hedrick AV, Kortet R (2006) Hiding behaviour in two cricket populations that differ in predation pressure. Anim Behav 72:1111–1118

Hill PSM, Wells H, Shadley JR (2006) Singing from a constructed burrow: why vary the shape of the burrow mouth? J Orthoptera Res 15(1):23–29

Houston RD, Boonman AM, Jones G (2004) Do echolocation signal parameters restrict bats' choice of prey? echolocation in bats and dolphins. University of Chicago Press, Chicago, pp 339–345

Ims RA (1990) On the adaptive value of reproductive synchrony as a predator-swamping strategy. Am Nat 136(4):485–498

Jacobs DS (1999) The diet of the insectivorous Hawaiian hoary bat (*Lasiurus cinereus semotus*) in an open and a cluttered habitat. Can J Zool 77(10):1603–1608

Jones G, Rydell J (2003) Attack and defense: interactions between echolocating bats and their insect prey. The University of Chicago Press, Chicago

Kalka M, Kalko EKV (2006) Gleaning bats as underestimated predators of herbivorous insects: diet of *Micronycteris microtis* (Phyllostomidae) in Panama. J Trop Ecol 22:1–10

Morrill SB, Fullard JH (1992) Auditory influences on the flight behavior of moths in a nearctic site flight tendency. Can J Zool 70(6):1097–1101

Moss CF, Zagaeski M (1994) Acoustic information available to bats using frequency-modulated sounds for the perception of insect prey. J Acoust Soc Amer 95:2745–2756

Nakano R, Skals N, Takanashi T, Surlykke A, Koike T, Yoshida K, Maruyama H, Tatsuki S, Ishikawa Y (2008) Moths produce extremely quite ultrasonic courtship songs by rubbing specialize scales. PNAS 105:11812–11817

Nakano R, Ishikawa Y, Tatsuki S, Skals N, Surlykke A, Takanahi T (2009) Private ultrasonic whispering in moths. Commun Integr Biol 2(2):123–126

O'Brien DM (1999) Fuel use in flight and its dependence on nectar feeding in the hawkmoth *Amphion floridensis*. J Exp Biol 202(4):441–451

Olofsson M, Vallin A, Jakobsson S, Wiklund C (2010) Marginal eyespots on butterfly wings deflect bird attacks under low light intensities with uv wavelengths. PLoS ONE 5(5):6

Pollack GS, Martins R (2007) Flight and hearing: ultrasound sensitivity differs between flight-capable and flight-incapable morphs of a wing-dimorphic cricket species. J Exp Biol 210 (18):3160–3164

Prestwich KN (1994) The energetics of acoustic signaling in anurans and insects. Am Zool 34 (6):625–643

Pulliam HR, Caraco T (1984) Living in groups: is there an optimal group size, vol 2. Sinauer, Sunderland Massachusetts

Ritzmann RE (1984) The cockroach escape response. Plenum Press, New York

Robinson MH (1969) Defenses against visually hunting predators. Evol Biol 3(22):5–59

Roeder KD (1998) Nerve cells and insect behavior. Harvard University Press, Cambridge

Roeder KD, Treat AE, VJ S (1968) Auditory sense in certain sphingid moths. Science 159 (3812):331–333

Rydell J (1998) Bat defence in lekking ghost swifts (*Hepialus humuli*), a moth without ultrasonic hearing. Proc Roy Soc Lond B Biol Sci 265(1404):1373–1376

Rydell J, Lancaster WC (2000) Flight and thermoregulation in moths were shaped by predation from bats. Oikos 88(1):13–18

Rydell J, Skals N, Surlykke A, Svensson M (1997) Hearing and bat defence in geometrid winter moths. Proc Roy Soc Lond B Biol Sci 264(1378):83–88

Schnitzler H-U, Kalko EKV (2001) Echolocation by insect-eating bats. Bioscience 51(7):557–569

Soutar AR, Fullard JH (2004) Nocturnal anti-predator adaptations in eared and earless Nearctic Lepidoptera. Behav Ecol 15(6):1016–1022

Spangler HG (1988) Hearing in tiger beetles (Cicindelidae). Physiol Entomol 13(4):447–452

Staudinger MD, Hanlon RT, Juanes F (2011) Primary and secondary defences of squid to cruising and ambush fish predators: variable tactics and their survival value. Anim Behav 81:585–594

Svensson AM, Danielsson I, Rydell J (2002) Avoidance of bats by water striders (*Aquarius najas*, Hemiptera). Hydrobiologia 489(1–3):83–90

Tauber E, Camhi JM (1995) The wind-evoked escape behavior of the cricket *Gryllus bimaculatus* —integration of behavioral elements. J Exp Biol 198(9):1895–1907

ter Hofstede HM, Killow J, Fullard JH (2009) Gleaning bat echolocation calls do not elicit antipredator behaviour in the Pacific field cricket, *Teleogryllus oceanicus* (Orthoptera: Gryllidae). J Comp Physiol A Neuroethology Sens Neural Behav Physiol 195(8):769–776

Triblehorn JD, Yager DD (2006) Wind generated by an attacking bat: anemometric measurements and detection by the praying mantis cercal system. J Exp Biol 209(8):1430–1440

Triblehorn JD, Ghose K, Bohn K, Moss CF, Yager DD (2008) Free-flight encounters between praying mantids (*Parasphendale agrionina*) and bats (*Eptesicus fuscus*). J Exp Biol 211 (4):555–562

Turchin P, Kareiva P (1989) Aggregation in *Aphis varians*—an effective strategy for reducing predation risk. Ecology 70(4):1008–1016

Watson CM, Roelke CE, Pasichnyk PN, Cox CL (2012) The fitness consequences of the autotomous blue tail in lizards: an empirical test of predator response using clay models. Zoology 115(5):339–344

Wicklein M, Strausfeld NJ (2000) Organization and significance of neurons that detect change of visual depth in the hawk moth Manduca sexta. J Comp Neurol 424(2):356–376

Yack JE (1988) Seasonal partitioning of atympanate moths in relation to bat activity. Can J Zool Rev Can Zool 66(3):753–755

Yager DD (1990) Sexual dimorphism of auditory function and structure in praying mantises (Mantodea; Dictyoptera). J Zool 221(4):517–537

Zeng J, Xiang N, Jiang L, Jones G, Zheng Y, Liu B, Zhang S (2011) Moth wing scales slightly increase absorbance of bat echolocation calls. PlosOne 6(11):e27190. doi:10.1371/journal. pone.0027190

Zuk M, Rotenberry JT, Tinghitella RM (2006) Silent night: adaptive disappearance of a sexual signal in a parasitized population of field crickets. Biol Lett 2(4):521–524

Chapter 4
Passive and Active Acoustic Defences of Prey Against Bat Predation

Abstract Perhaps the most effective and sophisticated defence evolved by insects in the context of bat predation is audition. Here, we review the evolution, ecology, and physiology of insect audition in the context of bat predation. Ears have evolved independently multiple times in insects and occur in almost half of extant moth species. In moths, these ears were used secondarily for intraspecific communication but in other orders (e.g. some crickets) the evolution of ears preceded bats and probably arose primarily for communication, the anti-bat function was a secondary. Insect ears are used as a primary defence mechanism to initiate a variety of acoustic startle responses involving the cessation of advertisement behaviours in response to bat echolocation. They are also used as a secondary defence mechanism to execute evasive manoeuvres after they have been detected by a bat. This requires that the insect is able to localize the bat. Localization of the predator by insect ears is achieved through difference in intensities amongst signals arriving at different ears. Bat predation has also influenced the advertisement behaviours of insects. For example, it is thought to be responsible for the evolution of tremulation and the complex songs in some insects. Finally, we provide a detailed review of the jamming, aposematic, and startle functions of moth clicks. A unique secondary defence in moths is the use of ultrasonic clicks to directly affect the foraging success of the bat. We discuss the three hypotheses advanced for the function of moth clicks—jamming, aposematism, and startle.

4.1 Evolution of Insect Ears

Echolocation enables bats to hunt insects successfully in complete darkness, and since its evolution bats have become the main predator of nocturnal insects. There are several hundred species of insectivorous bats, all of which use echolocation, to a lesser or greater extent while hunting, and most use ultrasonic calls. Bats use different modes of foraging (Chap. 2). Some bats attack insects in flight (aerial hawking), while others take stationary insects from the vegetation or ground (gleaning). Bats are not the only predators of insects that emit ultrasound. For

© The Author(s) 2016   43
D.S. Jacobs and A. Bastian, *Predator–Prey Interactions: Co-evolution between Bats and their Prey*, SpringerBriefs in Animal Sciences,
DOI 10.1007/978-3-319-32492-0_4

example, ultrasound is generated by carnivorous katydids, the calls of small insectivorous terrestrial mammals, and movement of these predators through vegetation (Robinson and Hall 2002). However, in comparison, echolocation is a conspicuous acoustic signal which can be readily exploited by prey to avoid predation. Such exploitation arose through the evolution of ears sensitive to the ultrasounds which comprise bat echolocation (Yack and Fullard 1993).

Despite the effectiveness of audition as an early warning system against an approaching bat, the only arthropods to have evolved audition as a defence against bats are insects. However, its effectiveness against bat predation is reflected in the independent evolution of audition at least 15–20 times in several different insect orders and lineages (Stumpner and von Helversen 2001; Strauß and Stumpner 2015).

Bat-detecting ears in insects consist of a thin cuticular membrane called the tympanum, with an air-filled tracheal cavity behind it. The tympanum is mechanically connected to a mechanosensitive chordotonal organ (Robert and Hoy 1998) which transmits vibrations of the tympanum to the auditory nerve. These tympanal organs are derived from proprioceptors (Yack and Fullard 1990) which are stretch receptors located all over the insect's body and allow the central nervous network to keep track of the relative position of different body parts. Derivation from such proprioceptors explains why insect ears are located in such odd places, including the abdomen, mouthparts, neck, thorax, and tibia (Hoy and Robert 1996; Yack 2004; Strauß and Stumpner 2015). Apart from the location of the proprioreceptors, the structure of insect ears (Fig. 4.1) suggests that other factors which influence their anatomical position include the following: (1) exposure to sound waves but also protection of the membrane by supporting structures (e.g. a chitinous ring in moths) that prevent non-acoustic vibrations from stimulating the tympanum; (2) the availability of an enlarged tracheal air sac (Fig. 4.1) that provides a pocket of air behind the tympanum for transmission of tympanal vibrations to the chordotonal organs on the interior surface of the tracheal sac and which connects to both ears enhancing the localization of the sound source in small insects despite the small

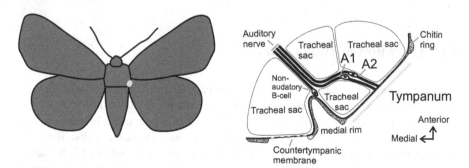

Fig. 4.1 Structure of an insect ear. A horizontal section through the right ear of a moth. Adapted from Yager (1999) (color figure online)

distance between the ears (Miles et al. 1995; Robert et al. 1996); (3) wide distance between the two ears in larger insects, assisting with the localization of sound, through differences in arrival times at each ear and the head or body shadow effect (Van Wanrooij and Van Opstal 2004) which causes differences in amplitude of the sound at each ear. Distances between ears must be smaller than half of one wavelength of the sound to facilitate sound localization (Miles et al. 1995; Robert et al. 1996; Mason et al. 2001; Ho and Narins 2006). Fortunately bats have to use high-frequency echolocation to locate objects the size of insects so the wavelength of their echolocation is small enough for most insect ears to meet this criterion (see Yack 2004 for a review of the structure and function of insect ears).

Tympanal hearing organs sensitive to ultrasound, for example in moths, have evolved multiple times in response to the evolution of echolocating bats and occur in almost half of extant species (Roeder and Treat 1957). The existence of ears in moths later enabled their use in communication and enabled the evolution of ultrasonic communication signals (Conner and Corcoran 2012; Greenfield 2014; Nakano et al. 2015a, b). The situation in other insects, e.g. some Orthoptera, may have been reversed in that ears may have evolved for acoustic communication first and then secondarily as defences against bats (Strauß and Stumpner 2015). Evidence for stridulation in older families of Orthoptera (e.g. Tettigonoidea) goes back about 250 million years (Gu et al. 2012), long before the advent of echolocating bats (65 million years ago; Eick et al. 2005; Teeling et al. 2005). Similarly, ears in mantids arose about 120 million years ago, before the advent of bats, and may have therefore also evolved for communication or in response to selection pressures from predators other than bats. Although mantids in flight do respond to attacking bats via evasive manoeuvres, there is currently no evidence for intraspecific acoustic communication in mantids (Strauß and Stumpner 2015). This suggests that mantid ears arose in response to predators other than bats.

Although the hearing range of moths is tuned to the echolocation frequencies of their sympatric bat communities (Fullard 1987; ter Hofstede et al. 2013), moths are unable to discriminate between different frequencies (Skals and Surlykke 2000; Fullard et al. 2007). Moth ears are simple (Fig. 4.1), containing between one and four sound-sensitive cells, nevertheless they provide moths with critical information about their acoustic environment (Miller and Surlykke 2001) and allow moths to detect the echolocation calls of most bats before the bats detect the echo reflected off of them (Roeder 1967; Goerlitz et al. 2010). For example, the Dogbane tiger moth (*Cycnia tenera*) does not respond to frequency shifts or to changes in the duration of bat echolocation calls. Instead, its ears are most sensitive to changes in the period and intensity of the bat calls (Fullard et al. 2007).

Not all insect ears are tone-deaf. It has been known for some time that locusts are able to discriminate different frequencies (Michelsen 1971), but with the advent of laser technology it has recently been discovered that both crickets and katydids are able to discriminate different frequencies (Montealegre-Z and Robert 2015). They do so by using a structure called the crista acustica, the insect equivalent of the mammalian basilar membrane (Stumpner and Nowotny 2014; Montealegre-Z et al. 2012). For a review of the basics of biomechanics of hearing in katydids, see Montealegre-Z and Robert (2015).

The dual function of insect ears and the behavioural context in which they have evolved have shaped the biomechanical properties of the ear. Ears which have evolved in the context of predation are highly sensitive to a broad range of acoustic frequencies (Pollack 2015), particularly in the ultrasonic range, and allow the localization of the sound. Reception of such acoustic signals results in rapid escape manoeuvres by the insects hearing them. In contrast, hearing in the context of communication requires the recognition and discrimination of highly specific patterns of sound (e.g. mating songs) as well as their localization. The latter requires directionality of hearing which is more precise than that required in the context of predator avoidance (Pollack 2015; Ronacher et al. 2015). Such localization can be achieved through side-specific gain control which typically allows the insect to encode the loudest signal on each side (see Pollack 2015 for a review).

Localization of the sound source in the context of predation can be achieved through difference in intensities amongst signals arriving at different ears. In this case, the sound shadow of the body is sufficient to cause a difference in the intensity of sound arriving at each ear. This allows the insect to determine the direction from which a bat is attacking and to execute evasive manoeuvres in the opposite direction. In contrast, the lower frequency sounds normally used in insect communication has longer wavelengths which are not interfered with by the insect's body. The difficulty in this case is localization of sounds that arrive almost simultaneously at both ears. These insects overcome this problem by using pressure-difference receivers in which the sound impinges on both the outer and the inner surfaces of the tympanum, and this membrane vibrates in response to the difference in pressure between the two sides (Montealegre-Z and Robert 2015).The filtered information is transmitted to the brain, where the final steps of pattern recognition and localization occur (Montealegre-Z and Robert 2015).

Insects that use hearing for both predator detection and mate attraction would benefit from the ability to segregate mating calls from those emitted by an attacking predator. This is achieved in some insects by the perceptual segregation of the mixture of the two kinds of signals reaching the ear into coherent representations of each of the signals. In the katydid, *Neoconocephalus retusus* (Orthoptera, Tettigoniidae), an auditory interneuron, segregates information about bat echolocation calls from background male advertisement songs using differences between the temporal and spectral characteristics of the two stimuli (Schul and Patterson 2003; Schul and Sheridan 2006).

Audition in insects make them as well adapted for detecting and avoiding bats as bats are for detecting and catching insects in a variety of different habitats and situations. Audition-based mechanisms of defence are as diverse as bat echolocation and can be divided into the same categories of primary and secondary defences (Staudinger et al. 2011) that are used to describe prey defences in general, but with the main difference being that they are based on sound.

4.2 Primary Acoustic Defence

4.2.1 Halting Reproductive Displays

Many insects will halt their reproductive displays when they become aware of an approaching bat. These responses are as varied amongst insects as the reproductive displays themselves.

4.2.1.1 Lepidoptera

Most moths use pheromones rather than sound to attract mates. Usually the females broadcast a pheromone plume that males then follow to the female. However, the use of pheromones alone does not protect moths from predation and quiet moths still have to employ additional strategies to avoid bats. Two species of moths, *Pseudaletia unipuncta* (Noctuidae) and *Ostrinia nubilalis* (Pyralidae), significantly reduced their mate-seeking behaviour in response to ultrasonic pulses which simulated high levels of predation risk by bats. In response to these simulated bat calls, males of these two species aborted their upwind flight in a pheromone plume and females stopped releasing pheromone (Acharya and McNeil 1998).

Although calling males in moths are not as common as in crickets, there are some moth species that use ultrasonic signals to attract females. Furthermore, these male moths will respond to acoustic stimuli which resemble the echolocation calls of bats, by ceasing to call. Such cessation of calling is also reported for stimuli that resemble the echolocation calls of gleaning bats (low amplitude, short duration of 1–3 ms, and very high frequency >60 kHz), thought to be inaudible to moths. However, male *Achroia grisella* moths will stop calling when exposed to such acoustic signals. This inhibition of male calling consists of two processes. First, there is a startle response to single short pulses of ultrasound during which male moths would miss broadcasting several calls. If subsequent pulses are delivered at a rate of 30 cycles per second, a silence response lasting for several seconds is elicited. Such inhibition is probably a specialized defensive behaviour that allows calling male moths to avoid detection by even very quiet bats that would use the male moths' calls to detect them (Greenfield and Baker 2003).

Similarly, male greater wax moths, *Galleria mellonella*, attract females over large distances using pheromones but also use short bursts of ultrasonic pulses for short distance communication (Skals and Surlykke 2000). The ultrasonic signals are probably used over short distances for greater ease of localization after the female has been attracted to the vicinity of the male by the pheromones (Skals and Surlykke 2000). However, there are other pyralids that use acoustic communication over long distances as well (Heller and Krahe 1994). Female *G. mellonella* respond by wing fanning which solicits mating opportunities from males. The problem is that the temporal and spectral characteristics of the males' calls are similar to the frequency-modulated echolocation used by bats to prey upon these moths. Female moths therefore need to distinguish bat echolocation calls from those of conspecific males so that they do not get eaten but also do not forego mating opportunities (Jones et al. 2002). Jones et al. (2002) showed that females could distinguish between the calls of predatory bats and those of conspecific males. They displayed by fanning their wings in response to conspecific male calls but not to bat calls. Furthermore, when receiving bat calls and male moth calls at the same time, females displayed more when the perceived risk was lower (search phase bat calls with lower call rate) and less when the perceived risk was higher (terminal phase bat calls with higher call rate). Their ability to discriminate between the two kinds of calls is based on the differences in the temporal characteristics of the calls (Jones et al. 2002; see also Greenfield and Baker 2003) rather than amplitude or frequency differences. There is no indication that these moths can discriminate different frequencies (Skals and Surlykke 2000). Instead, male song and bat echolocation are distinguished by the duration of the acoustic signal in the air and by the pulse rate on the ground in the moth *Achroia grisella*, the males of which do sing and attract females over great distances. In the air, any ultrasonic signal longer than 1 ms elicits a dive, i.e. the female moth regards it as an attacking bat (Rodriguez and Greenfield 2004). Here, amplitude does play a role because the response of female moths increased with increasing amplitude of the acoustic stimuli (Rodriguez and Greenfield 2004). On the substrate where moths are susceptible to attacks from gleaning bats, an ultrasonic signal repeated more than approximately 30 pulses per second is interpreted as a bat call (Greig and Greenfield 2004; Rodriguez and Greenfield 2004). There is much variation in the females' thresholds of response (Greig and Greenfield 2004; Brandt et al. 2005), but this may be due to the variation in the call rate of gleaning bats as they approach prey. Although some bats emit their search phase echolocation calls at between 10 and 30 calls per second (Neuweiler 2000) and do not increase this rate as they approach prey (Waters and Jones 1995), some gleaning species do increase their call rate as they approach their target (Swift and Racey 2002; Ratcliffe and Dawson 2003). A valuable field of research may therefore be to investigate female moth response to male moths and predatory bats in the context of their sympatric bat communities.

Curtailment of reproductive displays described in the above two examples may be tempered by competition between calling males for the attention of females such as in a lek situation. Such signal competition may make calling males more prone to taking risks, manifested in a reduction of the silence response. However, moths on leks also benefit from the dilution of predation pressure—the risk of attack on any particular individual is less in larger leks, and this may make males on leks call more. This was what was found in the pyralid moth, *Achroia grisella*. Male *A. grisella* gather in leks and broadcast ultrasonic mating calls. These males are preyed upon by substrate-gleaning bats, and calling males generally become silent when hearing bat echolocation calls. The incidence and duration of the silence response were greatly reduced in lekking males compared with solitary individuals when exposed to pulsed ultrasound with characteristics similar to those of bat echolocation calls. Furthermore, there was a small reduction in the silence response when males were first exposed to a song from a lek and then to synthetic bat echolocation calls. Thus, competition and dilution of predation pressure influences the silence response in moths (Brunel-Pons et al. 2011).

4.2.1.2   Orthoptera

Bats are known to feed extensively on Orthoptera. For example, crickets and katydids comprised 20% of the diet and was the 2nd most important dietary item, after moth caterpillars, of the gleaning phyllostomid, *Micronycteris microtis* (Kalka and Kalko 2006).

Foliage-gleaning bats often use echolocation for general orientation but tend to locate their prey passively by listening for the sounds they make. Four of the six katydid species taken by *Micronycteris hirsuta* on Barro Colorado Island in Panama call in the 23–27 kHz range, suggesting that bat foraging success is dependent on whether the bat can hear their calls or not (Belwood 1990). The selection pressure from bats has resulted in a number of adaptations in calling behaviour (Robinson and Hall 2002).

Many species of Orthoptera, like other nocturnal insects, have tympanal organs that are sensitive to ultrasound, e.g. in katydids (Tettigoniidae; Libersat and Hoy 1991), in grasshoppers and locusts (Acrididae; Robert 1989) and crickets (Gryllidae; Farris and Hoy 2000). The first evidence that orthopterans are responsive to ultra sound was found in mole crickets (Gryllotalpidae) by Suga (1968) who showed that two species of mole crickets respond to sounds with frequencies up to 70 kHz. The frequencies to which orthopteran ears are sensitive has a wide range from <20 kHz to as high as 100 kHz (McKay 1969, 1970; Kalmring and Kühne 1980). The ability to detect ultrasound enables Orthoptera to take evasive action to avoid being captured if a bat, or other sound-generating predator, approaches (Pollack 2015).

Orthoptera show a number of acoustic-induced startle responses (ASRs), including cessation or interruption of singing, cessation of flying, and flying away from the source of the sound (see review by Hoy 1992). Several species are known to stop singing in response to bats. The stridulating katydids *Conocephalus conocephalus* and *C. maculatus* in Nigeria and the Creosote bush katydid, *Insara covilleae*, of North America stop singing in the presence of flying bats (Sales and Pye 1974; Spangler 1984). *Insara covilleae* often calls from the top of bushes where it is vulnerable to gleaning bats. Its calls are above 20 kHz and within the hearing range of gleaning bats. This katydid will stop singing in response to the high-intensity calls of a bat that has approached close to its position but not during the low-intensity calls of a distant bat (Spangler 1984).

Audition in Orthoptera is also used in mate attraction, and it is usually the males that attract females by singing (Belwood and Morris 1987). In a reversal of the exploitation of bat echolocation by prey, the songs of male katydids are exploited by bats to locate prey (ter Hofstede et al. 2008; Jones et al. 2010; Chap. 2). Like calling moths, katydids have to discriminate between signals from bats and from conspecifics and behave appropriately if they do not want to get eaten or forego mating opportunities.

One Nearctic katydid species, *Neoconocephalus ensiger*, displays a silence response in response to an approaching bat. This was shown to be an effective defence against the sympatric gleaning bat *Myotis septentrionalis* which preferentially attacked a loudspeaker broadcasting *N. ensiger* songs over loudspeakers broadcasting a novel cricket song. Bats only attacked the speaker if the song was continuous and aborted their attack if there was a pause in the sound as the bats approached, regardless of whether a katydid was present as a physical target. Cessation of calling by the katydid was effective because these bats were using the songs of the katydid to locate it rather than echolocation. Echolocation used by the bats did not vary during approaches that involved continuous song or interrupted songs (ter Hofstede et al. 2008).

Although song cessation may be an effective defence against bats, the response of neotropical katydids preyed on by gleaning bats, which are known to use male calling songs to locate them, can vary depending on sensory and ecological factors. In Panamá, at least two katydid species (*Balboa tibialis* and *Ischnomela gracilis*, Pseudophyllinae) exposed to the echolocation calls of the gleaning phyllostomid, *Trachops cirrhosus*, stopped singing. These two species also had the highest number of spikes and firing rates in their T cells, an auditory interneuron sensitive to ultrasound. Another pseudophylline species, *Docidocercus gigliotosi*, did not demonstrate a consistent response to the bat calls and those that did respond only did so in response to playbacks of high intensity. *Docidocercus gigliotosi* also had a different T cell response curve to the bat call sequences, with spike rate decreasing with increasing intensity of the playbacks. Similarly, singing male tettigoniids, *Tettigonia cantans*, did not respond to the pulsed ultrasonic stimuli similar to the calls used by bats, whereas singing *T. viridissima* did when ultrasonic pulses occurred during the silent intervals in the male song or when the male was producing soft hemi-syllables (Hartbauer et al. 2010).

There is evidence that sensitivity to ultrasound has evolved in the context of bat predation. In some katydids (Tettigoniidae), the silence response is only initiated in response to ultrasound. Under controlled laboratory conditions, Faure and Hoy (2000) found that males cease calling or insert pauses in their songs only when stimulated with ultrasound (>20 kHz), but not when stimulated with audible sound (<20 kHz). Furthermore, this silence response to ultrasound only occurred when the stimulus arrived during the window of silence between stridulatory syllables. The simple explanation is that the katydid's own song, which is loud, temporarily deafens or masks its auditory system. Such suppression in some orthopterans during the production of stridulatory syllables (e.g. Wolf and von Helversen 1986; Hedwig 1990), may have resulted in the evolution of chirps and interrupted buzzes in male song. Such chirps and buzzes may provide "acoustic windows" during which singing males could monitor the songs not only of conspecifics (Greenfield 1990), but also of sounds made by predators (Faure and Hoy 2000).

This varied silence response to bat echolocation may be indicative of the different strategies used by different katydid species. Both *B. tibialis* and *I. gracilis* songs have significant energy at frequencies <15 kHz, whereas *D. gigliotosi* is purely ultrasonic (>20 kHz). Higher frequencies attenuate more rapidly in air than lower frequencies (Lawrence and Simmons 1982). The calls of *D. gigliotosi* might therefore be quieter than those of *B. tibialis* or *I. gracilis* over the same distance depending on the intensity at which these species emit their calls. Quieter songs could be a primary defence against gleaning bats allowing such katydids to sing for longer periods. However, some bat species, for example, *Lophostoma silvicolum*, another neotropical gleaning bat, locates the katydid, *Scopiorinus fragilis* (Tettigoniidae), using its songs which are at frequencies above >20 kHz (Belwood and Morris 1987). This suggests that higher frequencies alone are not an effective means of making katydid calls less audible to gleaning bats.

Some katydids may call at high duty cycle possibly as a result of competition between male katydids for females (Latimer and Sippel 1987; Bailey and Yeoh 1988; Bailey et al. 1990; Tuckerman et al. 1993), providing information about mate quality (Gwynne 2001). This may prevent them from hearing the echolocation calls of approaching bats. Instead of the silence response, such katydids might use alternative strategies. The copiphorine katydid, *N. affinis*, for example, does not stop singing in response to bat echolocation calls. Instead, it relies on a primary defence in the form of singing from habitat such as grasses and clearings, where sympatric neotropical gleaning bats are unlikely to forage (Belwood and Morris 1987; Kalko et al. 1999).

The absence of the silence response in some katydids may also be due to their auditory neuron being specialized for mate attraction. The phaneropterinae katydids, *Steirodon rufolineatum* (ter Hofstede et al. 2010) and *Amblycorypha oblongifolia* (ter Hofstede et al. 2008), also did not stop singing in response to bat echolocation calls. Female phaneropterinae respond to male calling by producing a tick-like sound between male calls (Heller and von Helversen 1986). The best sensitivity of the T cell in this group of crickets coincides with the peak frequency of the female call. In other katydid species, it centres around 20 kHz (ter Hofstede and Fullard 2008). Together, this evidence suggests that the T cell may function in mate-detecting for this group of katydids, and therefore is not available for bat detection, while these insects are on the ground (ter Hofstede et al. 2010).

Katydids are a highly diverse group with large variation in their songs, mating strategies, and ecology (Gwynne 2001). The diversity of katydids makes them a good system to investigate the evolution of sensory systems in the context of mate attraction and anti-predator behaviour. However, this has possibly resulted in an extreme taxonomic bias because relatively little is known about the passive acoustically mediated anti-predator behaviour of other orthopterans or other eared insects.

4.3 Secondary Acoustic Defence

4.3.1 Evasive Flight

4.3.1.1 Lepidoptera

ASRs in moths take the form of a variety of evasive manoeuvres from slow turns away from the source of the sound when it is at low intensity (Roeder 1967) to complex loops, spirals, and dives when the sounds are at higher intensities (Roeder 1962, 1975). Under field conditions when free-flying moths were exposed to a stationary source of ultrasonic pulses, the moths' responses appeared to be non-directional; in that it could be towards or away from the source of the sound but tended to move the moths towards the ground. Movement towards the ground resulted from a wider variety of flight manoeuvres, including passive and power dives, loops, rolls, and one or more tight turns (Fig. 4.2). In contrast, the responses of flying moths further from the source of the sound were directional. These moths experienced lower intensities because of their greater distance from the sound source and responded by turning and flying directly away from the sound source (Roeder 1962).

These varied responses to ultrasound of differing intensities are encoded by the period and interaural differences in the generation times of spikes in the acoustic cells of the moths. Inter-spike interval was important in eliciting diving behaviour in response to intense ultrasound, and the difference in spike generation between the ears on either side of the moths body allowed the directional response of turning

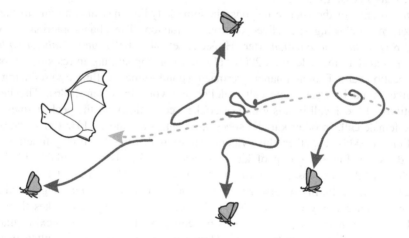

Fig. 4.2 Examples of evasive manoeuvres by insects in response to an aerial bat attack (color figure online)

away from the source of faint ultrasound (Roeder 1964). In noctuidae, such turning is affected by partial folding of the wings on the side of the moth opposite to the sound source (Roeder 1967).

4.3.1.2   Orthoptera

Many flying insects that are active at night effectively detect and evade bat predators through their sensitive ultrasonic hearing. Some eared insects can assess the level of risk by using the intensity of bat echolocation calls and the stereotyped changes in both amplitude and period of the calls as the bat moves through the different phases of its attack sequence, namely search, approach, and terminal phases. These eared insects can then mount effective anti-predator responses that vary from turning away from the predator (low risk of predation) to sudden random flight and dives (high risk of predation; Yager 2012).

Just as the silence response in insects are varied so too are their acoustic startle responses (ASR) to ultrasound during flight. The ground cricket, *Eunemobius carolinus* (Nemobiinae; Farris and Hoy 2000), and the tettigoniidae *Neoconocephalus ensiger* (Libersat and Hoy 1991; Faure and Hoy 2000), display ASRs in response to ultrasound while flying. This involves closing the wings, resulting in a free fall to the ground which takes the insects outside the cone of detection of the bat. Flight is stopped until the ultrasonic stimulus ends. *Neoconocephalus ensiger* also exhibits a non-flight ASR in the form of a silence response. Males stop singing in response to ultrasound arriving during the silent period between syllables, when the insects' own calls are paused.

Unlike *N. ensiger*, *Tettigonia viridissima* shows a directional response to simulated bat calls. It responds in one of the three ways, depending upon the intensity of ultrasound detected (Schulze and Schul 2001). When exposed to simulated bat calls at an intensity of 55–60 dB, the insect in tethered flight steered away from the sound source. At median intensities of 64 dB SPL, the insects stopped beating their hind wings but remained in the normal flight posture. Wing beats resumed 0.3–1 s after the simulated bat calls stopped. At median intensities of 76 dB SPL, *T. viridissima* folded its forewing into the resting position which would have resulted in a dive if the insect was untethered (Schulze and Schul 2001). The median threshold for the steering behaviour was 53.5 dB SPL, for the wing beat interruption 64 dB SPL and for diving 76 dB SPL.

Locusts (Robert 1989) and crickets (Lewis 1992) also show a directional response to ultrasound. Adult migratory locusts (*Locusta migratoria*: Acrididae) speed up their flight and turn away from ultrasonic pulses that mimic bat echolocation calls. This response is mediated by paired abdominal tympana and is considered an early warning system for bats. In fact indigenous farmers in Africa take advantage of the high-frequency hearing of locusts by rapping metallic pots to deflect swarms of locusts from their property (Robert 1989). Laboratory experiments on *Teleogryllus oceanicus* in flight have shown that it will turn towards the song of a conspecific and away from ultrasound (Moiseff et al. 1978; Nolen and Hoy 1986a). The switch between courtship and predator avoidance may be based on a difference in spike rate in the high-frequency ascending neuron Int 1 (ANA). Predator avoidance only occurs if activity in Int 1 is more than 180 spikes per second, while activity in response to courtship song is in the region of 35 spikes per second (Nolen and Hoy 1986b).

A note of caution with respect to experiments done on tethered insects is provided by the study of Dawson et al. (2004). Previous studies (Robert 1989; Dawson et al. 1997) have shown that tethered locusts display ASRs more often to trains of ultrasonic pulses (30 kHz) than to pulses below 10 kHz. These results were interpreted as an avoidance response to predatory bats. However, Dawson et al. (2004) investigated ASRs in locusts and moths, in both free flight and tethered, and found that turns loops and dives in locusts were initiated by a wide range of frequencies (high frequencies used by bats as well as low frequencies) and were statistically independent of the frequency of the stimulus. In contrast, ASRs in free-flying moths and tethered locusts were frequency-dependent and initiated by high frequencies similar to those used by bats (Dawson et al. 2004). Thus, although locust do have ASRs in free flight, tethering appears to accentuate their response to high-frequency acoustic stimuli (Dawson et al. 2004). It remains to be investigated why free-flying locusts display ASRs to such a wide range of frequencies (5–30 kHz). Possible explanations include flying locust being preyed upon by predators that produce acoustic signals that cover a wide range of frequencies such as birds (e.g. Jacobs et al. 2008) and bats (e.g. Fullard 1982).

For flight cessation to effectively drop an insect out of the echolocation cone of the bat, the insect must detect the bat before the bat can hear the echo of its call reflected off the insect. Despite hearing sensitivities of bats being generally higher than those of most tympanate insects (see Fullard 1987; Jacobs et al. 2008; Heffner et al. 2013), with the exception perhaps of those insects that have active auditory amplification (Mhatre 2015; Ronacher and Römer 2015), tympanate prey is able to hear most bats before the bats detect their echo. This is because insect prey listen to the primary signal, the bat call, whereas bats listen to the echo from this signal reflected off the prey. By the time the echo returns to the bat's ears, the energy initially in the bat's call has been attenuated by the atmosphere over twice the distance between the bat and the insect. The bat's call received by the insect has attenuated only over the distance between the bat and the prey. The intensity of the bat call at the insect's ears is therefore much higher than the intensity of the echo at the bat's ears. Schulze and Schul (2001) translated the intensities at which *T. viridissima* executed their different evasive manoeuvres (see above) into distance from the bat and found that steering behaviour would occur when the bat was 18 m away, interruption of the wing beat at 10 m, and the diving response at 5 m. In contrast, the bat would only detect the insect at a distance of 5 m (intensity of the bat echolocation taken as 110 dB SPL 25 cm in front of the bat).

This is also true for measurements of detection distance taken in the field. Schul et al. (2000) measured the hearing range of the long-winged katydid, *Phaneroptera falcata*, for echolocation calls of the mouse-eared bat, *Myotis myotis*, in the field while simultaneously monitoring the echolocation calls of the bats using two microphone arrays. This allowed them to reconstruct the flight path of the bat and determine the maximum distance at which the katydid detected the bat. They found that the katydid have hearing ranges of 13–30 m. They calculated that the katydid had more than 1 s to recognize the predator and respond evasively before the bat hears the returning echo. The hearing sensitivity of orthopterans, albeit generally lower than that of bats, and the evasive flight manoeuvres it allows them to execute are thus an effective defence against bat predation.

4.3.1.3   Mantodea

Like moths and katydids, mantids (Mantodea) have ears that are sensitive to ultrasound (Yager and Hoy 1989), allowing them to detect bat echolocation and to perform evasive manoeuvres to avoid capture (Yager et al. 1990). ASRs in mantids usually take the form of sudden turns (Yager and May 1990) or power dives to the ground taking the mantid out of the bat's cone of detection (Yager et al. 1990). Tethered male *Parasphendale agrionina* will not execute sudden turns in response to ultrasonic stimuli unless in flight. That this is an effective defence against attacks by echolocating bats is evidenced by the rate of survival of a bat attack of 42% lower in deafened mantids than in mantids with their ears intact (Triblehorn et al. 2008). Furthermore, in 25% of trials, power dives by the mantid forced the bat to completely abort its chase (25% trials) and in the remaining 75% of the trials, although the bats pursued diving mantids, the bats were nevertheless unsuccessful (Ghose et al. 2009). However, the rates of escape of mantids with intact ears decreased from 76 to <40% when pulse repetition rate in the bat's echolocation sequence increased rapidly. Therefore, an important difference between mantids and other insects with ASRs is that the ultrasound-sensitive interneuron likely involved in triggering mantis evasive responses shuts down during the last 200–300 ms of a bat attack (Triblehorn and Yager 2002). This is likely caused by the rapid transitions from low to high call rates when a bat homes in and captures an insect. Such rapid transitions potentially circumvent the mantid's ultrasonic defences, either by not triggering an evasive response or by triggering the response too late for the mantid to perform its evasive response (Yager and May 1990; Triblehorn and Yager 2005). This shortfall in auditory defences may be compensated for by the cerci of mantids (also found in other insects) that allow the detection of wind generated by the flapping wings of flying predators (see Chap. 3). The ASRs of mantids are, however, similar to those of moths (Fullard 1984) and lace wings (Miller 1975) in that they are very similar in their range of best frequencies. These ranges coincide with the frequency range of echolocation calls of the most common sympatric bats. However, moths are approximately 20 dB more sensitive to this range of frequencies than the other insects (Yager and May 1990).

4.4   The Form of Insect Calling Provides Evidence that These Behaviours Have Evolved in Response to Bat Predation

The calling behaviours of some orthopterans provide additional evidence that these behaviours are an evolutionary response to bat predation. Calling behaviour can vary amongst orthopterans in timing and location, and in some species can be replaced with complex species-specific vibrations called tremulations (Claridge 2006) to avoid signaling their presences to predatory bats. For example, katydids of the genera Aganacris, Scaphura, and Copiphora in central and southern America sing only when there is a lull in bat activity (Belwood 1990). At least one of these species, *Copiphora brevirostris*, is very common and palatable to captive bats, but is rarely taken by free-ranging gleaning bats (Belwood 1990). The strategy of calling when bats are less active is therefore an effective strategy against bat predation and probably evolved as a defensive strategy. Similarly, some neotropical katydids sympatric with foliage-gleaning bats call for a much smaller proportion of the time compared with species in areas where these bats are absent (Belwood and Morris 1987).

In most Pseudophyllus and Copiphora katydids, calling may be interrupted randomly with tremulations that can be detected by female conspecifics through the substrate but which cannot be detected by bats (Belwood and Morris 1987; Belwood 1990; Morris et al. 1994). Predation pressure from bats is the most likely reason for the evolution of tremulation and the switching from calls to tremulations (Morris et al. 1994). However, tremulation behaviour needs to be investigated in response to bat predation, possibly combined with phylogenetic information about when and in which lineages tremulation arose.

There is also evidence that the frequency and the components of katydid song may have been influenced by bat predation. Some species have calls with unusually high carrier frequencies. For example, the carrier frequencies of neotropical katydids ranged from 65 to 105 kHz (Morris et al. 1994). Atmospheric attenuation of such songs are high and only audible to bats over very short distances and therefore less likely to be heard by bats than songs of lower frequency. The most likely explanation out of a range of possible explanations was found to be bat predation (Morris et al. 1994).

Some species also have the ability to alter the components of their songs. For example, *Teleogryllus oceanicus* songs have a brief chirp followed by a trill. In laboratory experiments, males calling in the open and using calls of the same length were more likely to be attacked if their calls contained more trills than chirps (Bailey and Haythornthwaite 1998). In free-ranging katydids, males prefer calling from refuges but when calling from unprotected situations will use songs with fewer trills than chirps. Furthermore, unprotected insects were more likely to be attacked if they had longer calls with more elements (Bailey and Haythornthwaite 1998). That this calling behaviour is a response to bat predation is supported by the fact that the calls of neotropical katydids are shorter, with lower call rate and duty cycle and higher frequencies than that of temperate species where bat predation pressure is lower (Rentz 1975; Belwood 1990). Similarly, palaeotropical Malaysian leaf-mimicking katydids used calls of much lower frequency (<12 kHz) and higher duty cycle with more syllables than those of neotropical species. None of the Malaysian species appeared to use tremulation (Heller 1995). This is probably related to the lower predation pressure from bats in the palaeotropics, but it is nevertheless possible that it is a side effect of the different camouflage strategies adopted by the two groups of katydids (Heller 1995).

4.5   Moth Ultrasonic Clicks

In addition to their hearing defence, some moths in the subfamily Arctiinae also emit ultrasonic clicks of their own, and it was first suggested to have a defensive function against bat predation by Carpenter (1938) and Blest et al. (1963). Moths produce these clicks using tymbal organs located at the front of the thorax just behind the head. Tymbal organs are found on either side of the thorax of moths and consist of modified sclerites that are expanded and filled with air. On the upper anterior surface of the sclerite is a line of striations called the tymbal. Each striation or microtymbal is a shallow rectangular trough (Blest et al. 1963). Contraction of the muscles attached to these modified sclerites buckles the tymbals inward emitting a train of clicks. The number of clicks in each train is equivalent to the number of microtymbals and varies from one moth species to the next (Blest et al. 1963). When the muscles relax, the tymbal passively pops back to its original position emitting a second train of clicks as the microtymbals become unbuckled. In 1965, it was shown that bats would break off their attack on moths when the moths emitted their high-frequency clicks (Dunning and Roeder 1965), providing some of the first evidence that moth clicks were a defence against bats. Although much research has been done since then, we still do not know exactly how moth clicks deter bats. Three hypotheses, which may not be mutually exclusive, have been proposed (Miller and Surlykke 2001): the Jamming hypothesis (clicks jam bat echolocation by interfering with the bat's ability to process returning echoes), the Startle hypothesis (clicks startle the bat forcing it to abort its attack), and the Aposematic hypothesis (clicks alert the bat to the moths' unpalatability).

4.5.1   The Jamming Hypothesis

When hunting, bats can alter the spectral and temporal parameters of their echolocation calls to increase the rate at which it gathers information about the prey and the environment (Chap. 2). On the basis of these different phases of a bat's attack sequence, at least three hypotheses on the mechanism by means of which moths may jam bat echolocation have been advanced (Fullard et al. 1994; Corcoran et al. 2011): (1) the masking hypothesis (Møhl and Surlykke 1989; Troest and Møhl 1986), which proposes that moth clicks may mask the echoes from the moth's body making it temporally undetectable by the bat; (2) the phantom echo hypothesis (Fullard et al. 1979, 1994), which proposes that bats misinterpret moth clicks as echoes and the bat interprets the echoes as multiple objects forcing it to break off its attack; (3) the ranging interference hypothesis (Miller 1991; Masters and Raver 1996; Tougaard et al. 1998), which states that moth clicks compromises the bats' precision in determining the distance to the target reducing its capture success.

The masking hypothesis suggests that as soon as the moth clicks, the bat is no longer able to detect the moth and it should break off its attack immediately (unable to detect echoes from the moth. However, the ability of bats to continue tracking clicking prey suggests that moth clicks do not in fact mask echoes (Corcoran et al. 2011).

If moth clicks are interpreted by the bat as phantom echoes, the bat should respond to phantom echoes as if it was dealing with multiple targets. However, this is not the case. When exposed to multiple phantom echoes, bats did not avoid these multiple phantom targets or try to capture them, suggesting that they did not treat them as objects but as extrinsic sounds (Corcoran et al. 2011). Furthermore, this mechanism of jamming mechanism is based on the condition that moth clicks resemble the echoes from a bat's call in time–frequency structure and intensity. Since this condition is only met in the terminal phase (Chap. 2) of a bat's echolocation sequence during an attack, it has led to the prediction that moths should time their clicking to when the bat is in the terminal phase of its attack sequence (Fullard et al. 1994). This prediction was supported by the clicking behaviour of the arctiin moth, *Cycnia tenera*, in response to played back echolocation attack sequences of the big brown bat, *Eptesicus fuscus*. This moth clicked in the terminal phase of the bat's attack sequence (Fullard et al. 1994). However, support for this prediction was only found in *C. tenera*. Most other moths tested clicked well before the terminal phase of the bat's echolocation sequence (Barber and Conner 2006), and when exposed to phantom targets bats did not avoid or try to capture them suggesting that bats did not perceive these phantom echoes as objects (Corcoran et al. 2011).

The ranging interference hypothesis predicts that moth clicks should result in the bat perceiving the target as blurred compromising the precision with which the bat determines the distance to the moth (Corcoran et al. 2011). This appears to be the case. When big brown bats (*Eptesicus fuscus*) were allowed to attack clicking tiger moths, *Bertholdia trigona*, it frequently missed the moths by the same distance ($\sim$15–20 cm) predicted by the ranging interference hypothesis (Corcoran et al. 2011). Furthermore, sitting bats trained to determine whether a test echo was at the same distance as a reference echo, which was presented with or without recorded moth clicks, did not perform as well in this range discrimination test if the moth clicks were presented to the bat approx. 1.5 ms before the return of the echo (Miller 1991). Similar effects of moth clicks on a bat's ability to determine the distance to an object were also found by other studies (Masters and Raver 1996; Tougaard et al. 2003).

The timing of moth clicks for ranging interference: In the above studies, it was found that clicks which fell within a 2- to 3-ms window preceding the test signal had the greatest detrimental effect on the bat's ranging ability (Tougaard et al. 2003). Echoes from clutter (Chap. 2) had the greatest detrimental effect on a bat's ranging ability when it arrived within a window of 1–2 ms before the arrival of echoes from the target. Thus, if moth clicks were to jam a bat's echolocation calls it should arrive within a window of 1–3 ms. However, a moth needs between 25 and 35 ms (Fullard 1982) and between 80 and 150 ms (Fullard 1992) to produce clicks making it impossible for the moth to hear the first call of the bat and to place clicks within the 1- to 3-ms window before the return of the echo. It would also be impossible for the moth to predict when the bat will emit its next call to allow the moth to position its clicks appropriately because the temporal components of the echolocation sequence of an attacking bat are constantly changing. The best that a moth could hope to do is to produce as many clicks as possible during the terminal phase of a bat's attack sequence in the hope that at least some of the calls would be appropriately positioned (Barber and Conner 2006).

Jamming proposes that moths should click towards the end of a bat's attack sequence and that the higher the rate at which a moth produces its clicks the later in the bat's attack the moth should time its clicks (Barber and Conner 2006). The call rate of bats increases towards the end of the bat's attack sequence (Chap. 2). However, a survey of the clicking behaviour of several species of arctiin moths, including *Cycnia tenera*, showed that there was great variability in the timing of the clicks of different moth species in relation to the phase of a bats attack sequence. Most of the time, moth clicks did not coincide with terminal phase of the bat's attack sequence (Barber and Conner 2006). Furthermore, moths that produced their clicks at a higher rate did not have a greater tendency to click later in a bat's attack sequence than moths that clicked at a lower rate (Barber and Conner 2006). However, all of these studies involved moth species that clicked at relatively low rates. When jamming was tested with the tiger moth, *Bertholdia trigona*, which had a higher click rate, evidence for jamming was found (Corcoran et al. 2011). This moth produced ultrasonic clicks in response to bat attacks and was palatable producing no ill effects in bats that ate them. This meant that the refusal of bats to eat these moths was not due to the potential noxiousness of the moths but instead due to their clicks compromising the ranging performance of the bats.

In an elegant but simple experiment, a mixture of adult and young, naïve (to control for experience) bats were allowed to attack moths in a flight room (Corcoran et al. 2011). The bats were offered palatable moths that do not click, *B. trigona* that could click and *B. trigona* rendered clickless by the ablation of their tymbal organs. Bats completed their attacks on the non-clicking palatable moths about 400% more often than on clicking *B. trigona*. Furthermore, they attacked all *B. trigona* that could not click, suggesting that clicking was responsible for the bats not eating the clicking moths. The rate at which bats attacked clicking moths did not improve over time, suggesting that the bats did not habituate to the clicks as would be the case if the function of the clicks were to startle the bats. In addition, in persistent attacks on clicking *B. trigona*, bats altered their echolocation calls as if they were having difficulty locating the moths by echolocation, their echolocation apparently rendered ineffective by the moth clicks (Corcoran et al. 2011). These results provide convincing support for a jamming function for the moth clicks, at least in moths with high click rates, and suggest that by producing clicks at a high enough rate moths are able to ensure that at least some of their clicks have the correct temporal characteristics to interfere with bat echolocation.

However, this raises questions about the function of moth clicks produced at lower rates. A startle or aposematic function does not require that clicks have any particular temporal characteristics. The startle function only requires that clicking moths are rare and therefore rarely encountered by bats. This would ensure that they remain novel and that bats do not habituate to them. For maximum startle effect, the clicks should also be emitted just prior to the bat making contact, in much the same way that moths expose their eyespots to visually oriented predators at the last possible moment. In contrast, an aposematic function requires that clicks should always be associated with noxiousness and should be emitted early in the bat's attack sequence to give the bat time to process the message and to respond.

4.5.2 The Startle Hypothesis

Bats are known to be sensitive to unexpected sounds during the final moments of their attack sequence (Stoneman and Fenton 1988; Bates and Fenton 1990). Although naïve bats are startled by moth clicks, they quickly become habituated to clicking moths if the clicks are not associated with noxiousness (Bates and Fenton 1990; Miller 1991; Hristov and Conner 2005a; Corcoran et al. 2011). Bats habituated to moth clicks treated palatable clicking moths the same as palatable non-clicking moths (Hristov and Conner 2005a). However, for habituation to be maintained, bats would have to encounter moth clicks regularly. This may be unlikely if clicking moths are rare (<1% of the moth population—Bates and Fenton 1990). Most studies that have surveyed moth communities have found clicking moths in much higher proportions than this (Dunning et al. 1992), suggesting that moth clicks are unlikely to be effective as a means of startling bats.

4.5.3 The Aposematic Hypothesis

The aposematic hypothesis proposes that moth clicks may provide information to the bat about the noxiousness of the moth allowing the bat to abort its attack (Dunning and Roeder 1965; Dunning 1968). Aposematism is a defensive strategy by prey in which they accumulate toxins (usually obtained from their plant food) that render their body tissues unpalatable to predators. They then advertise this unpalatability by conspicuous traits, e.g. bright, contrasting colours for visually oriented predators (Rowe and Guilford 1999), which act as warning signals alerting the predator to their unpalatability. Acoustic aposematism is the use of an acoustic signal as a warning signal. Moth clicks may be a form of acoustic aposematism.

The intensity and temporal characteristics of bat echolocation calls can be used by eared moths to determine the distance and level of threat a bat represents. This allows the moth to temper its response accordingly. Unpalatable moths can click when the intensity and temporal characteristics of the bat's echolocation calls indicate that the bat has detected them. By doing so, they avoid attracting the attention of bats that had not detected them. Naïve bats learn to associate clicks with noxiousness through experience with noxious moths. However, bats have short memory retention and will forget the association between moth clicks and noxiousness after a short time has elapsed if the association is not reinforced (Bates and Fenton 1990). This association probably has to be re-learned often by bats if clicking moths are rare. Such learning places a high cost on moths because attacks from naïve bats can result in injury before the bats discover their unpalatability. The appropriate timing of clicking by the moth can avoid such injury. If a moth clicks when it is certain that it is under attack, it will have advertised its noxiousness without drawing unnecessary attention to itself. This would explain why most moth species with slow clicking rates click in the approach phase of a bat attack (Barber and Conner 2006; Ratcliffe and Fullard 2005). Clicking then gives the bat time to process the clicks and abort its attack. Clicking in the last moment of a bat's attack would minimize the risk of alerting a bat to the moth's presence but runs the risk of not giving the bat enough time to process the information and abort its attack.

Several moth species do in fact click in the last milliseconds of a bat attack. Some arctiin moths only click in response to tactile stimulation and not in response to bat echolocation calls (Barber and Conner 2006). Their clicks cannot therefore be used to jam the bats echolocation system and probably have an aposematic function. Bats often capture moths first in their wing or tail membranes before transferring the moths to their mouths. Clicking while still in the wing or tail can still communicate to the bat that the moth is unpalatable (Barber and Conner 2006). That this happens is supported by the fact that in moth collections it is mostly unpalatable species, rather than palatable ones, that have damaged wings (Rothschild 1985; Scoble 2002). Apparently unpalatable species that click are attacked and released, while palatable non-clicking moths are consumed (Brower 1984; Rothschild 1985; Scoble 2002).

Additional experimental support for the aposematic hypothesis comes from the manipulation of both palatability and the capacity to click. The aposematic hypothesis predicts that clicking should always be associated with noxiousness. This prediction was tested by comparing the rate at which naïve big brown bats captured tethered moths that differed in their levels of unpalatability and ability to click (Hristov and Conner 2005b). Capture rates were compared amongst species of moths that do not click, species of moths that click and are unpalatable and species of moths that normally click (and were unpalatable) but were rendered non-clicking by ablating the tymbal organs. The palatability of species that were normally noxious or palatable was varied by raising them on non-noxious and noxious plants, respectively. Naïve bats were allowed to learn about the association between unpalatability and clicking over 7 days, and then their response to different categories of moths was tested. Bats quickly learned to avoid clicking moths but only if the clicks were associated with a noxious taste. Similarly, Dunning (1968) and Bates and Fenton (1990) showed that bats were capable of associating clicks with bad taste.

Clicks do not have to be produced at a high rate to have a startle or aposematic function. The clicks have only to surprise the bat (startle) or be reliably associated with noxiousness (aposematic) and do not need to fall within the integration time of the bat's hearing (jamming). Clicks can startle bats but requires that clicking moths are rarely encountered by bats or that clicks are highly variable so that bats do not habituate to them. Currently, there is evidence in support of the aposematic function of moth clicks, at least in moth species that have a low rate of click emission. This is supported by the fact that the clicks of most arctiin species do not have the correct timing or temporal frequency structure to jam the echolocation systems of bats (Barber and Conner 2006; Ribeiro 2007; Corcoran et al. 2010). However, the encounter rates between bats and clicking/noxious moths used in these experiments must be similar to the encounter rates in the natural environment. This appears to be the case. The encounter rates of 25% used in Hristov and Conner's (2005b) experiments matched the proportions of clicking moths captured in light traps by Dunning et al. (1992). Similarly, about two-thirds of the moth species in a tropical arctiin assemblage had the ability to click and about 52% clicked in response to recorded bat echolocation calls (Barber and Conner 2006). However, bat–moth encounter rates are determined by the proportion of clicking moths in the entire moth assemblage that bats are exposed to, not just of the arctiin moths. Much lower proportions of clicking moths of the total moth population have been reported (less than 1%, Bates and Fenton 1990 and 2.3%, Ribeiro 2007). It would be informative to determine the rate of clicking by moths in different environments. If clicking moth proportions are low but the rate at which they click is high (clicking on >50% of the time in any particular time period), the function of clicking would be probably be to jam bat echolocation. If proportions and the rate of clicking are both low and associated with noxiousness, then aposematism is probably the best explanation for why moths click (Corcoran et al. 2010).

4.5.4 Moth Clicks: Variability and Function

If, as current the evidence suggests, that moth clicks either jams a bat's echolocation system or functions as an aposematic signal, natural selection should have reduced the variability of moth clicks. Reduction in variability would facilitate the jamming function of moth clicks by ensuring that they would increasingly have the appropriate temporal and spectral properties of bat calls in the terminal phase of the bats attack sequence. Similarly, if clicks evolved as an aposematic signal, the cost of predator learning on the prey population should exert selective pressure towards higher densities of clicking moths (i.e. clicking moths should not be so rare) and convergence in warning signals (i.e. clicks should not be variable). This would allow one species of moth to benefit from the predator learning induced by another species and Muellerian mimicry (see the later sect. in this chapter) should be more common. The fact that tiger moths cluster into two groups with clicks of different spectral and temporal characteristics, one set of characteristics suited for aposematism and the other for jamming (Corcoran et al. 2010), is strong evidence within cluster variation has been reduced by the selection for either an aposematism or jamming function. But can this explain the variability in moth clicks and does it totally preclude a startle function?

Moth clicks can vary within individuals, between individuals of the same species and between species (Conner 2007; Corcoran et al. 2010). This variability arises from differences in the number of microtymbals in each tymbal, changes in activation rates and duration of tymbal activity, and the degree of asynchrony between tymbals on either side of the body (Conner 2007). This is conducive to the startle hypothesis because the extreme variability of the clicks should prevent bats from becoming habituated to the clicks. Habituation in laboratory experiments (Bates and Fenton 1990; Miller 1991; Hristov and Conner 2005a) may thus be an artefact of experimental protocols that employed clicks with little or no variability.

Mathematical modelling of bat–moth interactions suggests that when clicking moths make up a small proportion of the moth population (2.3%), bats are unable to maintain the learned association between moth clicks and noxiousness (Conner 2007). However, this was dependent on the learning rates used in the model. At high learning rates, bats quickly learned to associate clicks with noxiousness and the numbers of naïve bats quickly declined. Under these conditions, the cost to the moths of bat learning was not high and there was only weak selective pressure for convergence of moth clicks. However, the proportion of clicking moths still needed to be relatively high to allow bats to learn the association between clicks and noxiousness (Barber and Conner 2007).

At the lower proportions (≤2.3%) evident in a natural population (Western Cape, South Africa), the startle function rather than the aposematic function received support (Ribeiro 2007). The startling effect was greatest when clicking variability is highest. If bats take only 3 trials to habituate to moth clicks (Bates and Fenton 1990), positive selection pressure is placed on unique click characteristics. This would imply that different species would remain with their specific characteristics and any variation within a population would be maintained through frequency-dependent selection. Frequency-dependent selection is commonly regarded as a mechanism by which variation in a particular trait is maintained in populations. Increased variation would prolong the time to habituation and may even prevent complete habituation to moth clicks within the bat population.

Another consequence of frequency-dependent selection is that it could, at least for a short time, increase the numbers of clicking moths in a population to the point where, if the moths are also noxious, the proportion of clicking moths could become high enough so that mortality and injury to moths as a result of bats having to learn the association between clicks and noxiousness is decreased, allowing the evolution of aposematism. Startle might therefore be a precursor function to aposematism. It might also explain why evidence has been found in support of both startle and aposematism (Bates and Fenton 1990; Miller 1991; Ribeiro 2007), why 20% of the naïve bats in Hristov and Conner's (2005a, b) study aborted their attacks on clicking arctiins despite these moths being palatable and why over the first two days of the interference trials in the Corcoran et al. (2011) study the bats appeared to be startled by moth clicks.

The proposals above assume that bats can differentiate between clicks and will only become habituated to clicks if they are similar. They also assume that bats are capable of making fine associations between click variability and palatability. Field work testing for these assumptions as well as the predictions of these models is needed, including work on the proportions of clicking moths in a variety of bat–moth communities, the rates at which bats become habituated to variable clicking and whether bats can identify different kinds of clicks.

It is likely that clicking has several functions (Corcoran et al. 2010) and which functions prevail in any given population may be dependent on the rate at which moths click, the variability of moth clicks, and the rarity of clicking moths in the prey community.

4.5.5   Acoustic Mimicry

Aposematism is energetically costly to the prey to produce both the noxiousness that renders them unpalatable and the warning signal that warns the predator about the noxiousness. Any moth that produces the warning signal but not noxiousness would be at an advantage over moths that produced both or just noxiousness. Noxiousness without a warning signal would still result in capture and injury before the bat discovers that the moth is noxious. However, producing only the warning signal may still incur the cost of injury as a result of a naïve predator attempting to consume it. An honest way of exploiting an aposematic system would be for different prey species that are all noxious to use the same warning signal, i.e. they mimic each other's signal. This increases the bat's encounter rate with the signal, it learns to associate the signal with noxiousness much quicker, and fewer moths are injured by naïve bats. This is a case of honest advertising by prey species in which the cost of predator learning is shared and all moths with the signal benefit. This is called Muellerian Mimicry (Müller 1879).

A deceptive way of exploiting the system is to parasitize it by producing the same warning signal but not the noxiousness thus foregoing costs associated with noxiousness and predator learning. This is called Batesian Mimicry. The palatable mimic benefits from the protection of the aposematic system but does not incur the cost of being noxious. Both forms of mimicry are well studied in systems involving visually oriented predators (e.g. birds) and prey that use colour and pattern (e.g. bright colours of caterpillars and butterflies) as warning signals.

Although there are anecdotal accounts of acoustic mimics in a variety of organisms (e.g. Dunning 1968; Young 2003), there are only a few confirmed examples of acoustic mimicry (e.g. Rowe et al. 1986) and only one example of mimicry in bat–moth interactions (Barber and Conner 2007). Barber and Conner (2007) provided evidence for both types of mimicry in bat–moth interactions. In a test for Muellerian Mimicry, bats were exposed in turn to palatable non-clicking noctuid moths as controls, noxious clicking tiger moths, *Cycnia tenera*, to a group of another noxious clicking moth, *Syntomeida epilais*, and finally to *S. epilais* that were rendered clickless by ablating their tymbals. In all trials, the palatable non-clicking moths were captured and eaten but the clicking and noxious *C. tenera* was not. When the clicking and noxious *S. epilais* were introduced, the bats avoided them suggesting that they had learned to associate the clicks with noxiousness and generalized this association to *S. epilais*. When non-clicking but noxious *S. epilais* were introduced the bats captured them but dropped them without consuming them. This suggests that the bats were associating the clicks with noxiousness and not some other cue, e.g. odour and that without clicks they were unable to identify noxious moths. Since both test moth species could click and were also unpalatable, this provided evidence for Muellerian Mimicry.

In a test of Batesian Mimicry, bats were again exposed to clicking and noxious *C. tenera* which were avoided. When exposed to a clicking palatable moth species, *Euchaetes egle*, bats took fewer of these than those of the palatable control moths, suggesting that the clicks protected *E. egle* despite its palatability. When the tymbals of *E. egle* moths were ablated so that they could no longer click bats captured and consumed them. Since *E. egle* provided the warning signal but not the noxiousness, this was evidence of deceptive acoustic mimicry, i.e. Batesian Mimicry.

The defences that insects have brought to bear against bat predation are truly remarkable, all the more so because they have originated from the modification of existing structures. However, their true magnificence is not so much in their diversity and effectiveness but in their simplicity.

References

Acharya L, McNeil JN (1998) Predation risk and mating behavior: the responses of moths to bat-like ultrasound. Behav Ecol 9(6):552–558

Bailey WJ, Haythornthwaite S (1998) Risks of calling by the field cricket *Teleogryllus oceanicus*; potential predation by Australian long-eared bats. J Zool 244(04):505–513

Bailey WJ, Yeoh PB (1988) Female phonotaxis and frequency discrimination in the bushcricket *Requena verticalis*. Physiol Entomol 13(4):363–372

Bailey WJ, Cunningham RJ, Lebel L (1990) Song power, spectral distribution and female phonotaxis in the bushcricket *Requena verticalis* (Tettigoniidae: Orthoptera): active female choice or passive attraction. Anim Behav 40(1):33–42

Barber JR, Conner WE (2006) Tiger moth responses to a simulated bat attack: timing and duty cycle. J Exp Biol 209(14):2637–2650

Barber JR, Conner WE (2007) Acoustic mimicry in a predator-prey interaction. Proc Natl Acad Sci USA 104(22):9331–9334

Bates DL, Fenton MB (1990) Aposematism or startle? predators learn their responses to the defenses of prey. Can J Zool 68(1):49–52

Belwood JJ (1990) Anti-predator defences and ecology of Neotropical forest katydids, especially the Pseudophyllinae. Crawford House Press, Bathurst, New South Wales, Australia

Belwood JJ, Morris GK (1987) Bat predation and its influence on calling behavior in Neotropical katydids. Science 238(4823):64–67

Blest AD, Collett TS, Pye JD (1963) The generation of ultrasonic signals by a new world arctiid moth. Proc Roy Soc Lond B Biol Sci 158(971):196–207

Brandt LRSE, Ludwar BC, Greenfield MD (2005) Co-occurrence of preference functions and acceptance thresholds in female choice: mate discrimination in the lesser wax moth. Ethology 111:609–625

Brower LP (1984) Chemical defence in butterflies. Paper presented at the symposia of the royal entomological society of London

Brunel-Pons O, Alem S, Greenfield MD (2011) The complex auditory scene at leks: balancing antipredator behaviour and competitive signalling in an acoustic moth. Anim Behav 81 (1):231–239

Carpenter GDH (1938) Audible emission of defensive froth by insects. Proc Zool Soc A 242–51

Claridge MF (2006) Insect sounds and communication: an introduction. In: Drosopoulos S, Claridge MF (eds) Insects sounds and communication: physiology, behaviour, ecology and evolution. Taylor & Francis Group, Boca Raton, pp 3–10

Conner D (2007) The defensive role of ultrasonic moth clicks against bat predation: a mathematical modeling approach. University of Cape Town, Cape Town. Retrieved from https://open.uct.ac.za/handle/11427/6172

Conner WE, Corcoran AJ (2012) Sound strategies: the 65-million-year-old battle between bats and insects. In: Berenbaum MR (ed) Annual review of entomology, vol 57. Annual Reviews, Palo Alto, pp 21–39

Corcoran AJ, Conner WE, Barber JR (2010) Anti-bat tiger moth sounds: form and function. Curr Zool 56(3):358–369

Corcoran AJ, Barber JR, Hristov NI, Conner WE (2011) How do tiger moths jam bat sonar? J Exp Biol 214(14):2416–2425

Dawson JW, Dawson-Scully K, Robert D, Robertson RM (1997) Forewing asymmetries during auditory avoidance in flying locusts. J Exp Biol 200(17):2323–2335

Dawson JW, Kutsch W, Robertson RM (2004) Auditory-evoked evasive manoeuvres in free-flying locusts and moths. J Comp Physiol A Neuroethology Sens Neural Behav Physiol 190(1):69–84

Dunning DC (1968) Warning sounds of moths. Zeitschrift für Tierpsychologie 25(2):129–138

Dunning DC, Roeder KD (1965) Moth sounds and insect-catching behavior of bats. Science 147 (3654):173–174

Dunning DC, Acharya L, Merriman CB, Dalferro L (1992) Interactions between bats and arctiid moths. Can J Zool Re Can Zool 70(11):2218–2223

Eick GN, Jacobs DS, Matthee CA (2005) A nuclear DNA phylogenetic perspective on the evolution of echolocation and historical biogeography of extant bats (Chiroptera). Mol Biol Evol 22(9):1869–1886

Farris HE, Hoy RR (2000) Ultrasound sensitivity in the cricket, *Eunemobius carolinus* (Gryllidae, Nemobiinae). J Acoust Soc Am 107(3):1727–1736

Faure PA, Hoy RR (2000) The sounds of silence: cessation of singing and song pausing are ultrasound-induced acoustic startle behaviors in the katydid *Neoconocephalus ensiger* (Orthoptera; Tettigoniidae). J Comp Physiol A Neuroethology Sens Neural Behav Physiol 186(2):129–142

Fullard JH (1982) Echolocation assemblages and their effects on moth auditory systems. Can J Zool 60(11):2572–2576

Fullard JH (1984) Listening for bats: pulse repetition rate as a cue for a defensive behavior in *Cycnia tenera* (Lepidoptera: Arctiidae). J Comp Physiol A 154:249–252

Fullard JH (1987) Sensory ecology and neuroethology of moths and bats: interaction in a global perspective. Cambridge University Press, Cambridge

Fullard JH (1992) The neuroethology of sound production in tiger moths (Lepidoptera, arctiidae) Rhythmicity and central control. J Comp Physiol A Neuroethology Sens Neural Behav Physiol 170(5):575–588

Fullard JH, Fenton MB, Simmons JA (1979) Jamming bat echolocation: the clicks of arctiid moths. Can J Zool 57:647–649

Fullard JH, Simmons JA, Saillant PA (1994) Jamming bat echolocation: the Dogbane tiger moth *Cycnia tenera* times its clicks to the terminal attack calls of the Big brown bat *Eptesicus fuscus*. J Exp Biol 194:285–298

Fullard JH, Ratcliffe JM, Christie CG (2007) Acoustic feature recognition in the dogbane tiger moth, *Cycnia tenera*. J Exp Biol 210(14):2481–2488

Ghose K, Triblehorn JD, Bohn K, Yager DD, Moss CF (2009) Behavioral responses of Big brown bats to dives by praying mantises. J Exp Biol 212(5):693–703

Goerlitz HR, Geberl C, Wiegrebe L (2010) Sonar detection of jittering real targets in a free-flying bat. J Acoust Soc Am 128(3):1467–1475

Greenfield MD (1990) Evolution of acoustic communication in the genus *Neoconocephalus*: discontinuous songs, synchrony, and interspecific interactions. In: Bailey WJ, Rentz DCF (eds) The tettigoniidae: biology, systematics and evolution. Springer, Berlin

Greenfield MD (2014) Acoustic communication in the nocturnal Lepidoptera insect hearing and acoustic communication. Springer, Berlin

Greenfield MD, Baker M (2003) Bat avoidance in non-aerial insects: the silence response of signaling males in an acoustic moth. Ethology 109(5):427–442

Greig EI, Greenfield MD (2004) Sexual selection and predator avoidance in an acoustic moth: discriminating females take fewer risks. Behaviour 141(7):799–815

Gu JJ, Montealegre-Z F, Robert D, Engel MS, Qiao GX, Ren D (2012) Wing stridulation in a Jurassic katydid (Insecta, Orthoptera) produced low-pitched musical calls to attract females. Proc Natl Acad Sci USA 109(10):3868–3873

Gwynne DT (2001) Katydids and bush-crickets: reproductive behavior and evolution of the Tettigoniidae. Cornell University Press

Hartbauer M, Ofner E, Grossauer V, Siemers BM (2010) The cercal organ may provide singing tettigoniids a backup sensory system for the detection of eavesdropping bats. PLoS ONE 5 (9):13

Hedwig B (1990) Modulation of auditory responsiveness in stridulating grasshoppers. J Comp Physiol A Neuroethology Sens Neural Behav Physiol 167(6):847–856

Heffner RS, Koay G, Heffner HE (2013) Hearing in American leaf-nosed bats. IV: the common vampire bat, *Desmodus rotundus*. Hear Res 296:42–50

Heller K-G (1995) Acoustic signalling in palaeotropical bushcrickets (Orthoptera: Tettigonioidea: Pseudophyllidea): does predation pressure by eavesdropping enemies differ in the Palaeo-and Neotropics? J Zool 237(3):469–485

Heller K-G, Krahe R (1994) Sound production and hearing in the pyralid moth *Symmoracma minoralis*. J Exp Biol 187:101–111

Heller K-G, von Helversen D (1986) Acoustic communication in phaneropterid bushcrickets: species-specific delay of female stridulatory response and matching male sensory time window. Behav Ecol Sociobiol 18(3):189–198

Ho CC, Narins PM (2006) Directionality of the pressure-difference receiver ears in the northern leopard frog, Rana pipiens pipiens. J Comp Physiol [A] 192(4):417–429. doi:10.1007/s00359-005-0080-7

Hoy RR (1992) The evolution of hearing in insects as an adaptation to predation from bats. In: Webster DB, Fay RR, Popper AN (eds) The evolutionary biology of hearing. Springer, New York, pp 115–129

Hoy RR, Robert D (1996) Tympanal hearing in insects. Annu Rev Entomol 41(1):433–450

Hristov NI, Conner WE (2005a) Sound strategy: acoustic aposematism in the bat–tiger moth arms race. Naturwissenschaften 92(4):164–169

Hristov N, Conner WE (2005b) Effectiveness of tiger moth (Lepidoptera, Arctiidae) chemical defenses against an insectivorous bat (*Eptesicus fuscus*). Chemoecology 15(2):105–113

Jacobs DS, Ratcliffe JM, Fullard JH (2008) Beware of bats, beware of birds: the auditory responses of eared moths to bat and bird predation. Behav Ecol 19(6):1333–1342

Jones G, Barabas A, Elliott W, Parsons S (2002) Female greater wax moths reduce sexual display behavior in relation to the potential risk of predation by echolocating bats. Behav Ecol 13 (3):375–380

Jones PL, Page RA, Hartbauer M, Siemers BM (2010) Behavioral evidence for eavesdropping on prey song in two Palearctic sibling bat species. Behav Ecol Sociobiol 65:333–340

Kalka M, Kalko EKV (2006) Gleaning bats as underestimated predators of herbivorous insects: diet of *Micronycteris microtis* (Phyllostomidae) in Panama. J Trop Ecol 22:1–10

Kalko EKV, Friemel D, Handley CO, Schnitzler H-U (1999) Roosting and foraging behavior of two Neotropical gleaning bats, *Tonatia silvicola* and *Trachops cirrhosus* (Phyllostomidae). Biotropica 31(2):344–353

Kalmring K, Kühne R (1980) The coding of airborne-sound and vibration signals in bimodal ventral-cord neurons of the grasshopper *Tettigonia cantans*. J Comp Physiol 139(4):267–275

Latimer W, Sippel M (1987) Acoustic cues for female choice and male competition in Tettigonia cantans. Anim Behav 35:887–900

Lawrence BD, Simmons JA (1982) Measurements of atmospheric attenuation at ultrasonic frequencies and the significance for echolocation by bats. J Acoust Soc Am 71(3):585–590

Lewis B (1992) The processing of auditory signals in the CNS of Orthoptera. In: Webster DB, Popper AN, Fay RR (eds) The evolutionary biology of hearing. Springer, New York, NY, pp 95–114

Libersat F, Hoy RR (1991) Ultrasonic startle behavior in bush-crickets (Orthoptera, tettigoniidae). J Comp Physiol A Neuroethology Sens Neural Behav Physiol 169(4):507–514

Mason AC, Oshinsky ML, Hoy RR (2001) Hyperacute directional hearing in a microscale auditory system. Nature 410(6829):686–690. doi:10.1038/35070564

Masters W, Raver K (1996) The degradation of distance discrimination in Big brown bats (*Eptesicus fuscus*) caused by different interference signals. J Comp Physiol A Neuroethology Sens Neural Behav Physiol 179(5):703–713

McKay JM (1969) The auditory system of Homorocoryphus (Tettigonioidea, Orthoptera). J Exp Biol 51(3):787–802

McKay JM (1970) Central control of an insect sensory interneurone. J Exp Biol 53(1):137–145

Mhatre N (2015) Active amplification in insect ears: mechanics, models and molecules. J Comp Physiol A 201:19–37. doi:10.1007/s00359-014-0969-0

Michelsen A (1971) Frequency sensitivity of single cells in the isolated ear the physiology of the locust ear (I-III). Springer, Berlin, pp 49–62

Miles RN, Robert D, Hoy RR (1995) Mechanically coupled ears for directional hearing in the parasitoid fly *Ormia ochracea*. J Acoust Soc Am 98(6):3059–3070. doi:10.1121/1.413830

Miller LA (1975) The behaviour of flying green lacewings, *Chrysopa carnea*, in the presence of ultrasound. J Insect Physiol 21(1):205–219

Miller LA (1991) Arctiid moth clicks can degrade the accuracy of range difference discrimination in echolocating big brown bats, *Eptesicus fuscus*. J Comp Physiol A Neuroethology Sens Neural Behav Physiol 168(5):571–579

Miller LA, Surlykke A (2001) How some insects detect and avoid being eaten by bats: tactics and countertactics of prey and predator. Bioscience 51(7):570–581

Møhl B, Surlykke A (1989) Detection of sonar signals in the presence of pulses of masking noise by the echolocating bat, *Eptesicus fuscus*. J Comp Physiol A 165:119–124

Moiseff A, Pollack GS, Hoy RR (1978) Steering responses of flying crickets to sound and ultrasound - mate attraction and predator avoidance. Proc Natl Acad Sci USA 75(8):4052–4056

Montealegre-Z F, Robert D (2015) Biomechanics of hearing in katydids. J Comp Physiol A Neuroethology Sens Neural Behav Physiol 201(1):5–18

Montealegre-Z ZF, Jonsson T, Robson-Brown KA, Postles M, Robert D (2012) Convergent evolution between insect and mammalian audition. Science 338(6109):968–971

Morris GK, Mason AC, Wall P, Belwood JJ (1994) High ultrasonic and tremulation signals in Neotropical katydids (Orthoptera, Tettigoniidae). J Zool 233:129–163

Müller F (1879) Ituna and Thyridia: a remarkable case of mimicry in butterflies. Trans Entomol Soc Lond 1879:20–29

Nakano R, Ihara F, Mishiro K, Toyama M, Toda S (2015a) High duty cycle pulses suppress orientation flights of crambid moths. J Insect Physiol 83:15–21

Nakano R, Takanashi T, Surlykke A (2015b) Moth hearing and sound communication. J Comp Physiol A Neuroethology Sens Neural Behav Physiol 201(1): 111–121

Neuweiler G (2000) The biology of bats. Oxford University Press, Oxford

Nolen TG, Hoy RR (1986a) Phonotaxis in flying crickets. Attraction to the calling song and avoidance of bat-like ultrasound are discrete behaviors. J Comp Physiol A Neuroethology Sens Neural Behav Physiol 159(4):423–439

Nolen TG, Hoy RR (1986b) Phonotaxis in flying crickets. Physiological-mechanisms of 2-tone suppression of the high-frequency avoidance steering behavior by the calling song. J Comp Physiol A Neuroethology Sens Neural Behav Physiol 159(4):441–456

Pollack GS (2015) Neurobiology of acoustically mediated predator detection. J Comp Physiol A Neuroethology Sens Neural Behav Physiol 201(1):99–109

Ratcliffe JM, Dawson JW (2003) Behavioural flexibility: the little brown bat, Myotis lucifugus, and the northern long-eared bat, M. septentrionalis, both glean and hawk prey. Anim Behav 66:847–856

Ratcliffe JM, Fullard JH (2005) The adaptive function of tiger moth clicks against echolocating bats: an experimental and synthetic approach. J Exp Biol 208(24):4689–4698

Rentz DC (1975) Two new katydids of the genus Melanonotus from Costa Rica with comments on their life history strategies (Tettigoniidae: Pseudophyllinae). Ent News 86:129–140

Ribeiro D (2007) The defensive role of ultrasonic moth clicks against bat predation: a mathematical modeling approach. University of Cape Town, Cape Town. Retrieved from https://open.uct.ac.za/handle/11427/6172

Robert D (1989) The auditory behaviour of flying locusts. J Exp Biol 147(1):279–301

Robert DR, Hoy RR (1998) The evolutionary innovation of tympanal hearing in Diptera comparative hearing: insects. Springer, pp. 197–227

Robert DR, Miles RN, Hoy RR (1996) Directional hearing by mechanical coupling in the parasitoid fly Ormia ochracea. J Comp Physiol [A] 179(1):29–44. doi:10.1007/BF00193432

Robinson DJ, Hall MJ (2002) Sound signalling in Orthoptera. In: Evans P (ed), Advances in insect physiology, vol 29. Academic Press, pp 151–278

Rodriguez RL, Greenfield MD (2004) Behavioural context regulates dual function of hearing in ultrasonic moths: bat avoidance and pair formation. Physiol Entomol 29:159–168

Roeder KD (1962) The behaviour of free flying moths in the presence of artificial ultrasonic pulses. Anim Behav 10(3):300–304

Roeder KD (1964) Aspects of the noctuid tympanic nerve response having significance in the avoidance of bats. J Insect Physiol 10(4):529–546

Roeder KD (1967) Turning tendency of moths exposed to ultrasound while in stationary flight. J Insect Physiol 13(6):873–888

Roeder KD (1975) Neural factors and evitability in insect behavior. J Exp Zool 194(1):75–88

Roeder KD, Treat AE (1957) Ultrasonic reception by the tympanic organ of noctuid moths. J Exp Zool 134(1):127–157

Ronacher B, Römer H (2015) Insect hearing: from physics to ecology. J Comp Physiol A Neuroethology Sens Neural Behav Physiol 201(1):1–4

Ronacher B, Hennig RM, Clemens J (2015) Computational principles underlying recognition of acoustic signals in grasshoppers and crickets. J Comp Physiol A Neuroethology Sens Neural Behav Physiol 201(1):61–71

Rothschild M (1985) British aposematic Lepidoptera. In: Heath J, Emmet AM (eds) The moths and butterflies of Great Britain and Ireland, vol 2. Harley Books, Colchester, UK, pp 9–62

Rowe C, Guilford T (1999) The evolution of multimodal warning displays. Evol Ecol 13 (7–8):655–671

Rowe MP, Coss RG, Owings DH (1986) Rattlesnake rattles and burrowing owl hisses: a case of acoustic Batesian mimicry. Ethology 72(1):53–71

Sales GD, Pye JD (1974) Ultrasonic communication by animals. Chapman & Hall, London

Schul J, Patterson AC (2003) What determines the tuning of hearing organs and the frequency of calls? a comparative study in the katydid genus *Neoconocephalus* (Orthoptera, Tettigoniidae). J Exp Biol 206(1):141–152

Schul J, Sheridan RA (2006) Auditory stream segregation in an insect. Neuroscience 138(1):1–4

Schul J, Matt F, von Helversen O (2000) Listening for bats: the hearing range of the bushcricket Phaneroptera falcata for bat echolocation calls measured in the field. Proc Roy Soc Lond B Biol Sci 267(1454):1711–1715

Schulze W, Schul J (2001) Ultrasound avoidance behaviour in the bushcricket *Tettigonia viridissima* (Orthoptera: Tettigoniidae). J Exp Biol 204(4):733–740

Scoble MJ (2002) The Lepidoptera. Form, function and diversity (Reprint 1992 ed). Oxford University Press

Skals N, Surlykke A (2000) Hearing and evasive behaviour in the greater wax moth, *Galleria mellonella* (Pyralidae). Physiol Entomol 25(4):354–362

Spangler HG (1984) Silence as a defense against predatory bats in two species of calling insects. Southwestern Nat 29(4):481–488

Staudinger MD, Hanlon RT, Juanes F (2011) Primary and secondary defences of squid to cruising and ambush fish predators: variable tactics and their survival value. Anim Behav 81:585–594

Stoneman M, Fenton M (1988) Disrupting foraging bats: the clicks of arctiid moths animal sonar. Springer, Berlin, pp 635–638

Strauß J, Stumpner A (2015) Selective forces on origin, adaptation and reduction of tympanal ears in insects. J Comp Physiol A Neuroethology Sens Neural Behav Physiol 201(1):155–169

Stumpner A, Nowotny M (2014) Neural processing in the bush-cricket auditory pathway insect hearing and acoustic communication. Springer, New York, pp 143–166

Stumpner A, von Helversen D (2001) Evolution and function of auditory systems in insects. Naturwissenschaften 88(4):159–170

Suga N (1968) Neural responses to sound in a Brazilian mole cricket. J Auditory Res 8(2):129–134

Swift SM, Racey PA (2002) Gleaning as a foraging strategy in Natterer's bat *Myotis nattereri*. Behav Ecol Sociobiol 52(5):408–416

Teeling EC, Springer MS, Madsen O, Bates P, O'Brien SJ, Murphy WJ (2005) A molecular phylogeny for bats illuminates biogeography and the fossil record. Science 307(5709):580–584

ter Hofstede HM, Fullard JH (2008) The neuroethology of song cessation in response to gleaning bat calls in two species of katydids, *Neoconocephalus ensiger* and *Amblycorypha oblongifolia*. J Exp Biol 211(15):2431–2441

ter Hofstede HM, Ratcliffe JM, Fullard JH (2008) The effectiveness of katydid (*Neoconocephalus ensiger*) song cessation as antipredator defence against the gleaning bat *Myotis septentrionalis*. Behav Ecol Sociobiol 63(2):217–226

ter Hofstede HM, Kalko EKV, Fullard JH (2010) Auditory-based defence against gleaning bats in Neotropical katydids (Orthoptera: Tettigoniidae). J Comp Physiol A Neuroethology Sens Neural Behav Physiol 196(5):349–358

ter Hofstede HM, Goerlitz HR, Ratcliffe JM, Holderied MW, Surlykke A (2013) The simple ears of noctuid moths are tuned to the calls of their sympatric bat community. J Exp Biol 216 (21):3954–3962

Tougaard J, Casseday J, Covey E (1998) Arctiid moths and bat echolocation: broad-band clicks interfere with neural responses to auditory stimuli in the nuclei of the lateral lemniscus of the big brown bat. J Comp Physiol A Neuroethology Sens Neural Behav Physiol 182(2):203–215

Tougaard J, Miller LA, Simmons JA (2003) The role of arctiid moth clicks in defense against echolocating bats: interference with temporal processing. In: Advances in the study of echolocation in bats and dolphins. University of Chicago Press, Chicago

Triblehorn JD, Yager DD (2002) Implanted electrode recordings from a praying mantis auditory interneuron during flying bat attacks. J Exp Biol 205(3):307–320

Triblehorn JD, Yager DD (2005) Timing of praying mantis evasive responses during simulated bat attack. J Exp Biol 208(10):1867–1876

Triblehorn JD, Ghose K, Bohn K, Moss CF, Yager DD (2008) Free-flight encounters between praying mantids (*Parasphendale agrionina*) and bats (*Eptesicus fuscus*). J Exp Biol 211 (4):555–562

Troest N, Møhl B (1986) The detection of phantom targets in noise by serotine bats; negative evidence for the coherent receiver. J Comp Physiol A 159:559–567

Tuckerman JF, Gwynne DT, Morris GK (1993) Reliable acoustic cues for female mate preference in a katydid (Scudderia curvicauda, Orthoptera, Tettigoniidae). Behav Ecol 4(2):106–113

Van Wanrooij MM, Van Opstal AJ (2004) Contribution of head shadow and pinna cues to chronic monaural sound localization. J Neurosci 24(17):4163–4171

Waters DA, Jones G (1995) Echolocation call structure and intensity in five species of insectivorous bats. J Exp Biol 198:475–489

Wolf H, von Helversen O (1986) 'Switching-off'of an auditory interneuron during stridulation in the acridid grasshopper *Chorthippus biguttulus* L. J Comp Physiol A 158(6):861–871

Yack JE (2004) The structure and function of auditory chordotonal organs in insects. Microsc Res Tech 63(6):315–337

Yack JE, Fullard JH (1990) The mechanoreceptive origin of insect tympanal organs: a comparative study of similar nerves in tympanate and atympanate moths. J Comp Neurol 300:523–534

Yack JE, Fullard JH (1993) What is an insect ear? Ann Entomol Soc Am 86(6):677–682

Yager DD (1999) Structure, development, and evolution of insect auditory systems. Microsc Res Tech 47:380–400

Yager DD (2012) Predator detection and evasion by flying insects. Curr Opin Neurobiol 22 (2):201–207

Yager DD, Hoy RR (1989) Audition in the praying mantis, *Mantis religiosa* L.: identification of an interneuron mediating ultrasonic hearing. J Comp Physiol A 165(4):471–493

Yager DD, May ML (1990) Ultrasound-triggered, flight-gated evasive maneuvers in the praying-mantis *Parasphendale agrionina*. Tethered flight. J Exp Biol 152:41–58

Yager DD, May ML, Fenton MB (1990) Ultrasound-triggered, flight-gated evasive maneuvers in the praying mantis Parasphendale agrionina. J Exp Biol 152:17–39

Young BA (2003) Snake bioacoustics: toward a richer understanding of the behavioral ecology of snakes. Q Rev Biol 78(3):303–325

References

Chapter 5
Aerial Warfare: Have Bats and Moths Co-evolved?

Abstract The interaction between bats and moths has been cited as an example of co-evolution. However, this is dependent on how well the diverse behavioural responses to prey detection and predator avoidance by bats and moths, respectively, satisfy the two major characteristics of co-evolution, specificity, and reciprocity. In general, co-evolution is an interaction between two species. Therefore, an interaction between multiple species of two orders as in the case of insects and bats may be a case of diffuse co-evolution. There is much evidence that moth anti-bat defences have evolved in direct response to bats. Such evidence includes the evolutionary origin of moth audition after bat echolocation, the close association between between moth hearing sensitivity and bat echolocation frequencies, the degeneration of hearing in moths that are no longer exposed to bat predation and the production of ultrasonic clicks by moths in direct response to bat echolocation. In contrast, evidence for reciprocity and specificity in the evolution of bat traits is confounded by the fact that these traits could also have evolved as adaptation for particular habitats and tasks. However, these requirements might be met by stealth echolocation, especially where these involve evolutionary trade-offs. For example, some bats use calls of low-intensity or low-frequency, sacrificing detection distance and the ability to detect small insects, respectively, that allow them to detect the moths before the moths detect them. Presumably, the decrease in detection distanced and the detectability is offset by an increased ability to catch eared/large moths.

5.1 Overview

The interaction between bats and their insect prey is one of the most cited examples of co-evolution (e.g. Fenton and Fullard 1979; Fullard 1988; Rydell et al. 1995; Miller and Surlykke 2001; ter Hofstede and Ratcliffe 2016). First suggested by Griffin in 1958 in his famous book *Listening in the Dark*, the evolutionary arms race between bats and moths, if it is an example of co-evolution, would have been initiated when bats evolved echolocation and moths evolved audition in response to

© The Author(s) 2016
D.S. Jacobs and A. Bastian, *Predator–Prey Interactions: Co-evolution between Bats and their Prey*, SpringerBriefs in Animal Sciences,
DOI 10.1007/978-3-319-32492-0_5

the selection pressure exerted on them by bat echolocation. Some bats in their turn supposedly improved their offensive weaponry by evolving stealth echolocation, i.e. echolocation at frequencies and intensities that are inaudible to moths. Several moth species already equipped with ears have also evolved ultrasonic clicks of their own that are used defensively against bat echolocation. This could be seen as the next step in the evolution of effective defences against bat echolocation. However, if the interaction between bats and moths is in fact an example of co-evolution, it has to satisfy the criteria of serial reciprocity and specificity (Futuyma and Slatkin 1983); i.e., moth defences must have evolved in response to bat echolocation and stealth echolocation must have evolved in response to moth hearing in the context of predation (Chap. 1). Since several species of bats probably interacted with several species of moths in this evolutionary arms race, it would probably be more accurate to describe any co-evolution between bats and moths as diffuse co-evolution; i.e. either or both lineages are represented by an array of species (Rothstein 1990). However, this does not preclude the possibility of as-yet undiscovered instances of pairwise co-evolution (Chap. 1) between bats and moths.

5.2   Moth Hearing as a Co-evolved Trait

5.2.1   Moth Hearing

Moths have apparently evolved ears with which to listen for the echolocation calls of approaching bats allowing them to respond to a bat attack by executing evasive flight manoeuvres. These manoeuvres include rapid flight in the opposite direction to the bat's approach if the bat is still far away, or if the bat is close, sharp dives up or down, sudden cessation of flight resulting in drops to the ground or flying in tight spirals inside the turning radius of the bat (Fig. 4.2). These manoeuvres avoid capture either because the bat cannot turn sharply enough to capture the moth and/or it takes the moth outside of the cone of detection of the bat's echolocation.

Moth ears probably evolved from stretch receptors (formerly known as proprioceptors), which are organs that allow the moth to keep track of the position of various parts of its body; i.e., if a leg or wing is folded close to its body or extended forward or backward away from the body (Yack 2004). This is very important information because it allows the moth to coordinate its movements. Understandably, such stretch receptors have to be positioned all over the moth's body to be effective. These stretch receptors were nothing more than the areas of thin cuticle innervated by a single nerve. Sometime in the evolutionary history of moths, a fortuitous mutation or mutations arose which resulted in the thinning of the cuticle that covers some of these stretch receptors. Areas of thin cuticle associated with a tracheal air sac would have formed an incipient ear, the thin cuticle acting as a tympanic membrane which could vibrate when sound pressure waves impinged upon it (Chap. 4). Such vibrations could have acted as an early warning system in moths carrying the mutation, resulting in them being slightly better at evading bats than moths without the mutation. This advantage would have resulted in better survival and reproduction of moths with the incipient ear. Because these ears arose independently in several lineages of moths and because they are the result of the evolutionary modification of stretch receptors positioned all over the moth's body, different species of moths have ears in some strange places including mouth parts, abdomen, thorax abdomen, and tibia (Hoy and Robert 1996; Yack 2004; Strauß and Stumpner 2015). Moth ears are most commonly placed on the thorax where it joins the abdomen of the moth, just behind the point of attachment of the wings. Less often, moth ears can occur on legs, mouthparts, and, in some crepuscular butterflies (the nymphalid, *Manataria maculata*), in the form of Vogel's organ at the base of the veins of the wings (Yack and Fullard 2000; Rydell et al. 2003). The incipient ears of moths would have represented an early warning system allowing moths with these ears to listen for the echolocation calls of an approaching bat and to take evasive action. In successive moth generations through several cycles of descent with modification, these ears became increasingly more sophisticated and effective in hearing bat echolocation calls allowing moths to be more successful at evading bats.

Moth audition has been very successful against some bats restricting them to eating insects other than moths or those few moth species that did not evolve ears. If moth hearing arose in response to bat predation, the evolutionary history of bats and moths should play out in tandem; i.e., moth ears should have evolved after bat echolocation. Unfortunately, very little work has been done on the history of moth hearing in the context of the co-evolution between bats and moths. However, there are some dated phylogenies emerging and when combined with historical bio-geography and fossil evidence for moths indicate that hearing in moths arose between about 60 and 70 mya (Yack and Fullard 2000; Grimaldi and Engel 2005). A possible exception is the evolution of hearing in the moth family Tineidea more than 90 mya (Davis 1998), although hearing at ultrasonic frequencies has not been confirmed in this group of moths (Greenfield 2016). Although bats arose during the same period as most eared moths, the emergence of bat echolocation apparently arose some time after that and is currently estimated to have arisen in the early Eocene (approximately 62–52 mya; Eick et al. 2005; Teeling et al. 2005; Simmons et al. 2008; Teeling 2009). A strict application of these dates suggests that the evolution of moth hearing may have preceded bat echolocation. However, there is much variation around these dates, and the fossil record, on which such dates are calibrated, is sparse for both bats and moths. Further data may very well result in greater coincidence of the timing of emergence of moth hearing and bat echolo-cation. In any case, the moth family Geometridae (c. 20,000 species), in which all lineages have ears with secondary loss of ears in only a few wingless females (Surlykke and Filskov 1997), arose about 46 mya (Yamamoto and Sota 2007). This places the origin of the Geometridae after the current estimated date of the origin of bat echolocation and within the time frame expected if hearing in this family of moth ears arose in response to bat predation. A further test of this prediction would be the compilation of more precisely dated phylogenies for both moths and bats.

The distribution of acoustic communication across the lepidopteran phylogeny suggests that sound-based intraspecific communication was preceded by the evolution of hearing which provided a sensory bias via which acoustic communication could evolve (Greenfield 2016). Tympanal organs occur in nine superfamilies (cf. Kristensen 2012; Regier et al. 2013), but the phylogenetic distribution of acoustic communication is more restricted and known primarily in the Pyraloidea, Papilionoidea, and Noctuoidea. Although distributed in multiple, unrelated genera and species, acoustic communication in these superfamilies is absent in most species (Greenfield 2016). The structures involved in sound production in these lineages are non-homologous and therefore probably had multiple independent evolutionary origins (Greenfield 2016). In contrast, the clicks produced by moths in response to the echolocation of an attacking bat probably arose as an anti-bat defence (Conner 1999) and then was secondarily co-opted for sexual communication (Spangler 1988; Nakano et al. 2009). Communication in the subfamily Arctiinae (Erebidae, formerly in the family Arctiidae; Zahiri et al. 2011) therefore evolved along different evolutionary pathways to that in other moth lineages (Conner and Corcoran 2012).

The paucity of data on the evolutionary history of moth hearing aside, there is much evidence suggesting that moth ears have evolved in response to bat echolocation. Firstly, ears have evolved in species that make no sounds of their own (i.e. ears in such species do not have a communicative role—Fullard 1987; Waters 2003). Ears have also evolved in nocturnal butterflies that have evolved from diurnal ancestors that do not have ears; i.e. when they became nocturnal, these species were exposed to bat predation and evolved ears as a defence against bat echolocation (Yack and Fullard 2000). Secondly, tympanate moths respond to bat echolocation by taking evasive action (Roeder 1967; Fullard 1982, 1987; Surlykke 1988; Miller and Surlykke 2001); i.e., their ears are functional in the context of bat predation. This response can be strong enough to interrupt mating behaviour in pyralid moths (Acharya and McNeil 1998). Thirdly, the ears of moths are most sensitive to frequencies used by the most common sympatric echolocating bats that prey on them (Fenton and Fullard 1979; Fullard 1982, 1987; Göpfert and Wasserthal 1999). In North America where most insect-eating bats echolocate between 20 and 50 kHz, this is also the range at which moths hear best (Fullard 1982, 1987, 1988). In Africa and Australia where moths are preyed upon by bats that use echolocation frequencies higher than 50 kHz, the upper hearing range of moths can extend to 100 kHz and beyond (Fenton and Fullard 1979; Fullard 1982, 1988; Fullard and Thomas 1981; Fullard et al. 2007; Jacobs et al. 2008). The ear of the noctuid moth *Prodenia* is sensitive to frequencies of up to 200 kHz (Roeder and Treat 1957).

The wide range of hearing sensitivities in moths also extends to lower frequencies (Fig. 5.1). Some moth species can hear frequencies below 20 kHz (Fig. 5.1) which allows them to hear the echolocation calls of bats that echolocate at low frequencies (e.g. *Otomops martiensseni* and *Tadarida aegyptiaca*; Fenton et al. 2002; Schoeman and Jacobs 2008). However, hearing sensitivity at low frequencies

Fig. 5.1 Auditory threshold curves for the A1 receptor in the noctuid, *Helicoverpa armigera*, and the notodontid, *Desmeocraera griseiviridis*. Redrawn from Jacobs 2016 (color figure online)

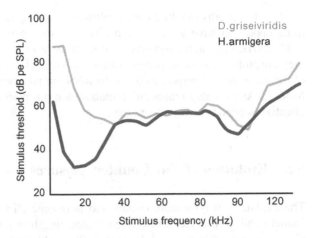

in moths may also be used to detect predatory birds (Jacobs et al. 2008). Such low-frequency sensitivity is relatively common in moths (Fig. 5.1; Fullard 1994; Rydell et al. 1995; Surlykke et al. 1998; Fullard et al. 2004; Jacobs et al. 2008) and may be associated with bird predation during the day, first suggested by Surlykke et al. (1998). Since birds use vision and are unlikely to sing while hunting moths, they offer very few acoustic cues to moths. However, when the birds have to alight on and probe through a bush to get at the moths, the rustling sounds they make may be used by moths as an early warning system against bird predation. There is evidence that moths do indeed use their hearing to listen to the rustling sounds made by predatory birds as they move through vegetation hunting moths (Jacobs et al. 2008).

Lastly, moths not exposed to bat predation, but which, by virtue of being tympanate, presumably were exposed to such threats in their evolutionary past, appear to undergo auditory degeneration. Moths that occur in areas where bats are absent (e.g. on isolated oceanic islands) have reduced auditory sensitivity to high frequencies but have retained auditory sensitivity to low frequencies (Fullard 1994; Fullard et al. 2004) and, more importantly, have lost the acoustic startle response to bat echolocation calls shown by moths that are sympatric with bats. Similarly, diurnal moths that have evolved from nocturnal eared ancestors have lost their ability to hear high-frequency sound (Fullard et al. 1997; Surlykke et al. 1998). In their new diurnal habitat, they are no longer exposed to bats, most of which are nocturnal. Sensitivity to ultrasound is therefore no longer an advantage and regresses, e.g. the diurnal geometrid *Archiearis parthenias* (Surlykke et al. 1998). Conversely, butterflies in the family Hedyloidea that have secondarily become nocturnal have gained hearing (Yack et al. 2007). Auditory degeneration is also evident within the same species among flightless individuals which are less exposed than volant individuals to aerially foraging bats. For example, flightless female gypsy moths (*Lymantria dispar*) have reduced hearing capacity compared to their volant male counterparts (Cardone and Fullard 1988).

Very few moths use their ears or ultrasonic signals in communication, and these traits appear to have arisen specifically in response to bat echolocation (Fullard et al. 2004). Although co-evolved moth clicks probably arose after moth audition, they are probably not a response to a counter-adaptation in bat echolocation but one of the many diverse responses of moths to the initial emergence of bat echolocation. Moth clicks probably represent an elaboration on moth defences allowing a more effective secondary anti-bat strategy.

5.3 Evolution of Bat Countermeasures

The evolution of tympanate organs and ultrasonic clicks in moths and their subsequent ability to detect and evade echolocating bats influences the foraging efficiency of bats (Fenton and Fullard 1979). Although this is expected to exert selection pressure on bat echolocation, bats also have to deal with a range of selection pressures from their environment and the challenges imposed upon them by the physical laws of sound propagation. These pressures may limit their ability to respond to the anti-bat defences of moth. However, large differences in the consumption of moths among bats, in general, and the consumption of eared moths by some species of bats suggest that some bats (e.g. the high-frequency echolocators in the families Hipposideridae, Rhinolophidae, and Vespertiliondae and a few low-frequency echolocators in the Molossidae) may have evolved such countermeasures making them better at catching moths (Jones 1992; Jacobs 2000; Schoeman and Jacobs 2003). This in turn suggests that the traits and strategies that allow these bats to hunt moths successfully may have evolved in direct response to the evolution of moth hearing. If so, it may fulfil the specificity and reciprocity criteria of co-evolution.

Bats may reduce the acoustic conspicuousness of their echolocation calls to eared moths by using frequencies outside of the hearing range of moths, reducing the intensity of their calls and the duration of their calls, or a combination of all three. Evidence that some species of bats may have evolved echolocation frequencies that are inaudible to their insect prey comes in the form of support for the allotonic frequency hypothesis (AFH—Novick 1977; Fenton and Fullard 1979). The AFH proposes that bats have reduced their acoustic conspicuousness by evolving echolocation frequencies that fall outside the range of frequencies (hence allotonic—a tone other than that that can be heard by insect prey) at which moths hear best. This has been traditionally regarded as between 20 and 50 kHz (e.g. Fullard 1987; Fullard and Dawson 1997; Bogdanowicz et al. 1999), the range at which the most common bats echolocate (ter Hofstede and Ratcliffe 2016). However, this range can be broader than this depending on the specific range of frequencies used by the sympatric bat community to which moths are exposed (Fullard and Thomas 1981; Fullard 1987; Jacobs et al. 2008). Thus, relatively high or low echolocation frequencies in bats may be the result of selection that has made some bat species less audible to eared insects, especially moths. If so, species using allotonic frequencies should prey heavily on eared insects, while those using syntonic frequencies (frequencies which fall within the hearing ranges of insect prey) should be less successful at capturing eared insects. This appears to be the case for several species of molossid and at least one vespertilionid. For example, the spotted bat, *Euderma maculatum* (Vespertilionidae), which uses anomalously low-frequency calls (12 kHz—Woodsworth et al. 1981; Fenton and Bell 1981) for its weight (Jones 1996, 1999; ter Hofstede and Ratcliffe 2016), outside of the range of sympatric eared moths (Fullard and Dawson 1997) fed predominantly on moths. *Euderma maculatum* ate both tympanate (Erebidae, Geometridae) and atympanate (Lasiocampidae) moth species at greater proportions than existed in its habitat (Painter et al. 2009). Similarly, the molossid, the Large-eared giant mastiff bat (*Otomops martiensseni*) also uses anomalously low frequencies (11.8 ± 0.3 kHz, Schoeman and Jacobs 2008) for its body weight. At approximately 30 g, it should call at a frequency of about 35 kHz (Jones 1996, 1999). The calls of *Otomops* may thus be allotonic with respect to its sympatric moth prey which could explain why it feeds predominantly on moths (Rydell and Yalden 1997). Unfortunately, whether this and other molossids using low-frequency echolocation feed predominantly on tympanate moths is unknown. However, these molossids and *E. maculatum* are all fast-flying, open-space aerial hawkers (Norberg and Rayner 1987; Rydell and Yalden 1997), and their low-frequency calls are less affected by atmospheric attenuation allowing these bats to detect large insects at relatively longer distances (Rydell and Yalden 1997; reviewed by Ratcliffe 2009). The adaptive advantage of such low-frequency echolocation calls may be increased detection distance rather than decreased audibility to eared insects (ter Hofstede and Ratcliffe 2016). Furthermore, many moths have sensitive hearing at frequencies below 20 kHz (Fig. 5.1, see also Fullard 1987). Although this has been ascribed to bird predation (Jacobs et al. 2008) or an adaptation allowing the detection of low-frequency social calls used by bats (Fullard 1987), frequencies used by these bats may not in fact be allotonic to their sympatric moth communities.

At the other end of the frequency range of bat echolocation are species such as the rhinonycterid (formerly in the family Hipposideridae, see Foley et al. 2014), *Cloeotis percivali*, with unusually high call frequency (208 ± 2.5 kHz, Jacobs 2000), which preys almost exclusively on moths (Jacobs 2000; Schoeman and Jacobs 2011). Furthermore, it has been suggested that the high-frequency calls of bats in the families Rhinolophidae and Hipposideridae have been selected because they are less audible to tympanate moths (Fenton and Fullard 1979; Fullard 1987; Jones 1992). Certainly, in the communities in which these bats occur, there is a correlation between call frequency and the proportion of moths consumed, and species in these families have both the highest call frequency and consume the most moths (Jacobs 2000; Schoeman and Jacobs 2003). In fact, the relationship between call frequency and the amount of moths consumed has been found at different levels of analyses including among different families of bats (Jones 1992; Bogdanowicz et al. 1999), different communities of bats (Jacobs 2000; Schoeman and Jacobs 2003; Schoeman and Jacobs 2008), and individual species (Pavey and Burwell 1998).

However, such correlations on their own cannot be used as definitive support for the AFH. Firstly, the moths being eaten by bats supposedly using allotonic call frequencies must be eared moths, and one cannot determine this from a simple examination of the proportion of moth remains in the faeces of bats because bats tend to masticate their insect prey very finely or discard identifiable fragments before ingesting (but see Pavey and Burwell 1998). Although ears are ubiquitous across moth species and most moths in any natural community are likely to be eared, an adequate test of whether allotonic frequencies in bats are an evolutionary response to the evolution of ears in moths requires that the identity of the moth remains in bat faeces be known so that one may determine whether the moths consumed are in fact tympanate. Fortunately, recent developments in genetic sequencing will allow researchers to use DNA from insect remains in bat faeces to reliably identify the moth species eaten by bats. Unfortunately, such analyses have not been employed in the study of the diets of bats in the three families of HDC bats, Hipposideridae, Rhinolophidae, and Rhinonycteridae.

Another factor confounding the ascription of allotonic frequencies to co-evolution with moth hearing is that such frequencies may have evolved in response to other factors. For example, the high frequencies used by rhinolophids, rhinonycteris, and hipposiderids may have evolved in response to their habitat characteristics rather than moth hearing. The echolocation calls of HDC bats are uniquely adapted for flutter target detection in highly cluttered habitats (Neuweiler 1980; Schuller 1984; Neuweiler 1989). The calls of HDC species combine high frequencies with long durations. These long-duration calls (>10 ms; Jones and Rayner 1989) are adaptations that allow bats to detect fluttering prey by the acoustic glints generated when the echolocation pulse is reflected off the wings of fluttering insects at different angles during the wing beat cycle (Chap. 2; Schnitzler and Flieger 1983; von der Emde and Schnitzler 1986). If so, the evolution of high-frequency and long-duration calls evolved because it allows the generation of such acoustic glints off the wings of even tiny insects allowing HDC bats to distinguish the echoes from prey from background echoes; i.e., these calls are clutter-resistant. The calls of most LDC bats are not clutter-resistant. Furthermore, because moths tend to have relatively larger wings than other insects they may appear more abundant to HDC bats because their larger wings generate acoustic glints more readily. In combination, these factors may explain the higher preponderance of moths in the diets of HDC bats. The reciprocity criterion of co-evolution may not therefore be met.

Another prediction of co-evolution and the AFH is that the calls of HDC bats should be inaudible to moths. However, the echolocation calls of some of these bats which echolocate at frequencies between 50 and 100 kHz are audible to sympatric moths (Fullard et al. 2008; Jacobs et al. 2008) as a result of their long duration (Jacobs et al. 2008) and because sympatric moths appear to have hearing sensitivities which match the frequencies of the echolocation calls used by these bats (Fullard et al. 2008; Jacobs et al. 2008). For example, two common South African moth species, *Helicoverpa armigera* (Noctuidae) and *Desmeocraera griseiviridis* (Notodontidae; Fig. 5.1), and the Australian Granny's cloak moth *Speiredonia spectans* (Noctuidae, Fullard et al. 2008) had increased hearing sensitivities of between 80 and 100 kHz, which spanned the echolocation frequencies of sympatric rhinolophids (Fullard et al. 2008; Jacobs et al. 2008). Most of the calls of the Dusky leaf-nosed bat, *Hipposideros ater* (150–160 kHz), were however inaudible to *S. spectans*. Nevertheless, the frequency-modulated portion of the calls of this bat, which were slightly lower in frequency but still in excess of 120 kHz, was audible to *S. spectans*. However, only 16% of these calls elicited action potentials in the auditory nerve of the moth (Fullard et al. 2008). Thus, although the calls of some hipposiderids may be functionally inaudible to moths, some moths are capable of hearing at frequencies above 100 kHz. This suggests that moths are sensitive to even the higher echolocation frequencies used by bats and that these frequencies are not therefore allotonic.

The evolution of high frequencies used by most HDC bats may be a consequence of their habitat and the evolution of an echolocation system that relies on Doppler shifts to deal with the acoustic challenges of these habitats (Waters 2003). As far as is known, all HDC bats hunt close to or within vegetation (Heller and von Helversen 1989). In these circumstances, detection distances of bats by moths are likely to be short, giving moths both less time and less space within which to react to an attacking bat. This makes moths more vulnerable to bat predation. Given that rhinolophids are thought to have originated in forest habitats in equatorial regions of the Old World, a more parsimonious explanation is that habitat was the major driver of echolocation call frequency in the Rhinolophidae with the colonization of less cluttered habitats by species appearing later in the lineage, resulting in calls of lower frequency (Stoffberg et al. 2011; Dool et al. 2016).

Alternatively, or in addition, the frequencies used by HDC bats may be a consequence of Doppler shift compensation (DSC, Fig. 2.1). The clutter-resistant echolocation system of HDC bats is based on DSC (Chap. 2). Since Doppler shifts are less pronounced at lower flight speeds but more pronounced at higher frequencies, smaller HDC bats flying at lower speeds may have evolved higher frequencies to increase Doppler shifts. This may facilitate DSC because the acoustic fovea can be defined by a wider range of frequencies and the bats can use less precise shifts in frequency (Waters 2003). This may be advantageous in the generation and detection of glints from small insects. If so, high frequencies in HDC bats may not have evolved in response to insect hearing.

However, bats may have responded to moth hearing by reducing the intensity and duration of their calls, making their calls less audible to moths (Fullard 1987). Short calls (<6 ms) of low intensity have the effect of reducing the distance at which moths can reliably detect bats giving moths less time to execute evasive manoeuvres (Fullard 1987; Waters and Jones 1996). The short duration and lower intensity calls used by some gleaning bats may therefore provide a means of circumventing moth hearing (Fenton and Fullard 1979; Fullard 1992; Waters and Jones 1996). Examples of gleaning bats whose calls are relatively inaudible to moths include *Myotis evotis* (Faure et al. 1990) and *M. septentrionalis* (Faure et al. 1993). There is evidence that the eared arctiin moth, *Cycnia tenera*, and noctuid moths with more sensitive ears cannot hear the echolocation calls of the gleaning bat *M. septentrionalis* (Faure et al. 1993; Ratcliffe and Fullard 2005). This moth usually emits ultrasonic clicks when it hears the echolocation calls of an attacking bat. When exposed to *M. septentrionalis,* it did not produce calls in response to the echolocation calls of this bat and only clicked in response to being handled by this bat (Ratcliffe and Fullard 2005). However, although low-intensity calls seemingly have a selective advantage in being less audible to moths, this advantage is probably secondary. It is unlikely that such calls evolved because they were less detectable by moths. Low-intensity calls of short duration are suited to the foraging habitat and mode used by substrate gleaning bats. Such calls minimize backward masking effects (Chap. 2, Fig. 2.2) allowing the bats to discriminate prey from clutter, increase resolution, and prevent self-deafening (Chap. 2, Fenton et al. 1995). This kind of echolocation probably evolved within this context because it allowed these bats to avoid pulse-echo overlap when foraging close to the ground and vegetation. This may especially be so since the calls of these bats form an adaptive complex with their wings which allows these bats to negotiate the tight spaces that gleaning bats have to deal with while gleaning prey from the ground and vegetation (Schnitzler and Kalko 2001).

Evolutionary trade-offs may make it easier to unambiguously identify evidence for bat counter-adaptations. Evolutionary trade-offs in the context of bat-prey interactions are adaptations in bat echolocation that are advantages because they make bat calls less audible to moths but that result in some disadvantage to the bat, e.g. disruption of the relationship between call frequency and wing shape or a decrease in resolution or detection range. Presumably the advantage of increased ability to catch and eat moths more than offsets the disadvantage of, for example, decreased detection distance. This may have occurred in the evolution of the aerial-hawking insectivorous barbastelle bat, *Barbastella barbastellus* (Goerlitz et al. 2010a, b). The intensity at which the echolocation calls of this species is emitted is much lower (10–100 times lower) than that of other aerial-hawking bats. The disadvantage of such low-intensity calls is that the detection range of the bats is much shorter. However, moths are unable to hear these low-intensity calls until the bat is too close for the moth to execute effective evasive manoeuvres (Goerlitz et al. 2010a). The effectiveness of this strategy in overcoming the hearing defence of moths is supported by the fact that *B. barbastellus* feeds almost entirely on moths (Vaughan 1997). Of greater importance, examination of the diet of this bat using DNA to identify the species of insect remains in the faeces of this bat has shown that this bat feeds almost exclusively on eared moths (Goerlitz et al. 2010a).

Many substrate gleaning bats are less conspicuous to moths because they use echolocation calls of low amplitude that are not audible to moths (e.g. *Myotis evotis* and *M. septentrionalis,* Faure et al. 1990, 1993) and/or passive listening to locate prey (e.g. *M. myotis* and *M. blythii,* Russo et al. 2007). However, it is not clear that these strategies have evolved to make these bats less conspicuous to moths. Low-intensity calls may have evolved to minimize the masking of target echoes by background echoes (Schnitzler and Kalko 2001). It may also minimize the masking of prey-generated sounds used by the bat to passively locate prey by the echolocation calls of the bat (Russo et al. 2007).

Another tantalizing potential response by bats to moth hearing is the two-tone echolocation calls of some bats. The Velvety free-tailed bat (*Molossus molossus*) uses echolocation that consists of a sequence of pairs of calls that differ in frequency (Mora et al. 2014). The first call in the pair of calls has a frequency of 34.5 kHz and the second call a slightly higher frequency at 39.6 kHz. Although such frequency alternation has been shown to increase the ranging ability and the amount of time that bat is continuously scanning its environment, it also makes the calls of this bat less audible to eared moths. Flying moths are known to use intensity and repetition rates of bat calls to determine whether the bat has initiated an attack on it before executing avoidance manoeuvres (Roeder 1964; Corcoran et al. 2013). An increase in call intensity and a decrease in call interval would convey to the moth that a bat is approaching it. A two-tone call interferes with the moth's processing of changes in call intensity and call interval because the second call in the pair of bat calls falls within a less sensitive region of the moths hearing range and has to be at a higher intensity for the moth to hear it. As a consequence, the moth does not detect the second call and interprets the call interval as being much longer than it actually is and, therefore, that the bat much further away than it actually is. This will result in the moth not executing evasive manoeuvres timeously and increases its probability of capture by the bat (Mora et al. 2014). However, whether these two-tone calls arose in direct response to moth hearing, and is therefore an example of co-evolution, is not known. An alternative explanation is that such calls evolved because it increases both the maximum detection distance, by parallel processing of call and echo in two different frequency channels, and the duty cycle of the bat's echolocation system. This would allow the bat to scan for insect prey with greater temporal continuity (Mora et al. 2014; Jung et al. 2007).

Similar reservations can be advanced with respect to calls of bats in the family Mormoopidae which use multiple harmonic calls. The first harmonic is syntonic with moth hearing and of low intensity, and the higher harmonics are allotonic to moth calls which should allow them to effectively circumvent moth hearing (Mora et al. 2013). However, these calls may also have arisen as adaptations to the bats' foraging habitat (Chap. 2).

Ultrasonic clicks for jamming or as part of an aposematic system are extremely effective anti-bat strategies employed by moths and are expected to exert selection pressure on bats to evolve countermeasures. However, there is no evidence that the evolution of bats has responded reciprocally or specifically to moth clicks.

In conclusion, it is likely that the interactions between at least some bat species and the moths they prey upon are good examples of an evolutionary arms race. However, the bat's response is likely to be more complex than just a shift in frequency as suggested by the allotonic frequency hypothesis. Given the recent advances in technology, investigation of the contributions of both low-intensity and allotonic frequencies to the evolution of stealth echolocation in bats should be undertaken. In addition, surveys of moth auditory capabilities should be broadened to improve both geographic and taxonomic coverage. Lastly, it is essential to tease apart the relative influences on the evolution of bat echolocation of both prey defences and the acoustic environments in which bats operate.

References

Acharya L, McNeil JN (1998) Predation risk and mating behavior: the responses of moths to bat-like ultrasound. Behav Ecol 9(6):552–558

Bogdanowicz W, Fenton M, Daleszczyk K (1999) The relationships between echolocation calls, morphology and diet in insectivorous bats. J Zool 247(03):381–393

Cardone B, Fullard JH (1988) Auditory characteristics and sexual dimorphism in the gypsy-moth. Physiol Entomol 13(1):9–14

Conner WE (1999) Un chant d'appel amoureux: acoustic communication in moths. J Exp Biol 202:1711–1723

Conner WE, Corcoran AJ (2012) Sound strategies: the 65-million-year-old battle between bats and insects. In: Berenbaum MR (ed) Ann Rev Entomol, vol 57. Annual Reviews, Palo Alto, pp 21–39

Corcoran AJ, Wagner RD, Conner WE (2013) Optimal predator risk assessment by the sonar-jamming arctiine moth *Bertholdia trigona*. PLoS ONE 8(5):13

Davis DR (1998) A world classification of the Harmacloninae, a new subfamily of Tineidae (Lepidoptera: Tineoidea). Smithson Contrib Zool 597:1–57

Dool SE, Puechmaille SJ, Foley NM, Allegrini B, Bastian A, Mutumi GL, Maluleke TG, Odendaal LJ, Teeling EC, Jacobs DS (2016) Nuclear introns outperform mitochondrial DNA in inter-specific phylogenetic reconstruction: lessons from horseshoe bats (Rhinolophidae: Chiroptera). Mol Phylogenet Evol 97:196–212

Eick GN, Jacobs DS, Matthee CA (2005) A nuclear DNA phylogenetic perspective on the evolution of echolocation and historical biogeography of extant bats (Chiroptera). Mol Biol Evol 22(9):1869–1886

Faure PA, Fullard JH, Barclay RMR (1990) The response of tympanate moths to the echolocation calls of a substrate gleaning bat, *Myotis evotis*. J Comp Physiol A: Neuroethol Sens Neural Behav Physiol 166(6):843–849

Faure PA, Fullard JH, Dawson JW (1993) The gleaning attacks of the northern long-eared bat, *Myotis septentrionalis*, are relatively inaudible to moths. J Exp Biol 178:173–189

Fenton MB, Fullard JH (1979) The influence of moth hearing on bat echolocation strategies. J Comp Physiol 132(1):77–86

Fenton MB, Audet D, Obrist MK, Rydell J (1995) Signal strength, timing, and self-deafening - the evolution of echolocation in bats. Paleobiology 21(2):229–242

Fenton MB, Bell GP (1981) Recognition of species of insectivorous bats by their echolocation calls. J Mammal 62(2):233–243

Fenton MB, Taylor PJ, Jacobs DS, Richardson EJ, Bernard E, Bouchard S, Debaeremaeker KR, ter Hofstede H, Hollis L, Lausen CL, Lister JS, Rambaldini D, Ratcliffe JM, Reddy E (2002) Researching little-known species: the African bat *Otomops martiensseni* (Chiroptera: Molossidae). Biodivers Conserv 11(9):1583–1606

Foley NM, Thong VD, Soisook P, Goodman SM, Armstrong KN, Jacobs DS, Puechmaille SJ, Teeling EC (2014) How and why overcome the impediments to resolution: lessons from rhinolophid and hipposiderid bats. Mol Biol Evol 32(2):313–333

Fullard JH (1982) Echolocation assemblages and their effects on moth auditory systems. Can J Zool 60(11):2572–2576

Fullard JH (1987) Sensory ecology and neuroethology of moths and bats: interaction in a global perspective. Cambridge University Press, Cambridge

Fullard JH (1988) The tuning of moth ears. Experientia 44(5):423–428

Fullard JH (1992) The neuroethology of sound production in tiger moths (Lepidoptera, arctiidae) Rhythmicity and central control. J Comp Physiol A: Neuroethol Sens Neural Behav Physiol 170(5):575–588

Fullard JH (1994) Auditory changes in noctuid moths endemic to a bat-free habitat. J Evol Biol 7(4):435–445

Fullard JH, Dawson JW (1997) The echolocation calls of the spotted bat *Euderma maculatum* are relatively inaudible to moths. J Exp Biol 200(1):129–137

Fullard JH, Thomas DW (1981) Detection of certain African, insectivorous bats by sympatric, tympanate moths. J Comp Physiol 143(3):363–368

Fullard JH, Dawson JW, Otero LD, Surlykke A (1997) Bat-deafness in day-flying moths (Lepidoptera, Notodontidae, Dioptinae). J Comp Physiol A: Neuroethol Sens Neural Behav Physiol 181(5):477–483

Fullard JH, Ratcliffe JM, Soutar AR (2004) Extinction of the acoustic startle response in moths endemic to a bat-free habitat. J Evol Biol 17(4):856–861

Fullard JH, Ratcliffe JM, Christie CG (2007) Acoustic feature recognition in the dogbane tiger moth, *Cycnia tenera*. J Exp Biol 210(14):2481–2488

Fullard JH, Jackson ME, Jacobs DS, Pavey CR, Burwell CJ (2008) Surviving cave bats: auditory and behavioural defences in the Australian noctuid moth, *Speiredonia spectans*. J Exp Biol 211:3808–3815. doi:10.1242/jeb.023978

Futuyma DJ, Slatkin M (1983) Introduction. In: Coevolution. Sunderland. Sinauer Associates Inc, Massachusetts

Goerlitz HR, ter Hofstede HM, Zeale MR, Jones G, Holderied MW (2010a) An aerial-hawking bat uses stealth echolocation to counter moth hearing. Curr Biol 20(17):1568–1572

Goerlitz HR, Geberl C, Wiegrebe L (2010b) Sonar detection of jittering real targets in a free-flying bat. J Acoust Soc Am 128(3):1467–1475

Göpfert MC, Wasserthal LT (1999) Hearing with the mouthparts: behavioural responses and the structural basis of ultrasound perception in acherontiine hawkmoths. J Exp Biol 202(8):909–918

Greenfield MD (2016) Evolution of acoustic communication in insects. In GS Pollack, AC Mason, AN Popper, RR Fay (eds) Insect hearing. Springer Handbook of Auditory Research, vol 55, pp 17–47

Griffin DR (1958) Listening in the dark: the acoustic orientation of bats and men. Yale University Press, New Haven

Grimaldi D, Engel MS (2005) Evolution of the Insects. Cambridge University Press, Cambridge

Heller K-G, von Helversen D (1989) Resource partitioning of sonar frequency bands in rhinolophoid bats. Oecologia 80(2):178–186

Hoy RR, Robert D (1996) Tympanal hearing in insects. Annu Rev Entomol 41(1):433–450

Jacobs DS (2000) Community level support for the allotonic frequency hypothesis. Acta Chiropterol 2(2):197–207

Jacobs DS (2016) Evolution's chimera: bats and the marvel of evolutionary adaptation. University of Cape Town Press, Cape Town

Jacobs DS, Ratcliffe JM, Fullard JH (2008) Beware of bats, beware of birds: the auditory responses of eared moths to bat and bird predation. Behav Ecol 19(6):1333–1342

Jones G (1992) Bats vs moths: studies on the diets of rhinolophid and hipposiderid bats support the allotonic frequency hypothesis. Prague stud Mammal 87–92

Jones G (1996) Does echolocation constrain the evolution of body size in bats? Symp Zool Soc Lond 69:111–128

Jones G (1999) Scaling of echolocation call parameters in bats. J Exp Biol 202:3359–3367

Jones G, Rayner JM (1989) Foraging behavior and echolocation of wild horseshoe bats *Rhinolophus ferrumequinum* and *R. hipposideros* (Chiroptera, Rhinolophidae). Behav Ecol Sociobiol 25(3):183–191

Jung K, Kalko EK, von Helversen O (2007) Echolocation calls in Central American emballonurid bats: signal design and call frequency alternation. J Zool 272(2):125–137

Kristensen NP (2012) Molecular phylogenies, morphological homologies and the evolution of moth "ears". Syst Entomol 37:237–239

Miller L, Surlykke A (2001) How some insects detect and avoid being eaten by bats. Bioscience 51(7):570–581

Mora EC, Macías S, Hechavarría J, Vater M, Kössl M (2013) Evolution of the heteroharmonic strategy for target-range computation in the echolocation of Mormoopidae. Front Physiol 4

Mora EC, Fernández Y, Hechavarría J, Pérez M (2014) Tone-deaf ears in moths may limit the acoustic detection of two-tone bats. Brain Behav Evol 83(4):275–285

Nakano R, Ishikawa Y, Tatsuki S, Skals N, Surlykke A, Takanahi T (2009) Private ultrasonic whispering in moths. Commun Integr Biol 2(2):123–126

Neuweiler G (1980) How bats detect flying insects. Phys Today 33(8):34–40

Neuweiler G (1989) Foraging ecology and audition in echolocating bats. Trends Ecol Evol 4 (6):160–166

Norberg UM, Rayner JMV (1987) Ecological morphology and flight in bats (Mammalia; Chiroptera): wing adaptations, flight performance, foraging strategy and echolocation. Philosophical Trans Royal Soc, Lond B 316:335–427

Novick A (1977) Acoustic orientation. In WA Wimsatt (ed) Biology of bats. Academic Press, New York and London, vol 3, pp 73–287

Painter ML, Chambers CL, Siders M, Doucett RR, Whitaker JO Jr, Phillips DL (2009) Diet of spotted bats (*Euderma maculatum*) in Arizona as indicated by fecal analysis and stable isotopes. Can J Zool 87:865–875

Pavey CR, Burwell CJ (1998) Bat predation on eared moths: a test of the allotonic frequency hypothesis. Oikos 143–51

Ratcliffe JM (2009) 11 predator-prey interaction in an auditory world. In: Dukas R, Ratcliffe J (eds) Cognitive ecology II. University of Chicago Press, Chicago, pp 201–228

Ratcliffe JM, Fullard JH (2005) The adaptive function of tiger moth clicks against echolocating bats: an experimental and synthetic approach. J Exp Biol 208(24):4689–4698

Regier JC, Mitter C, Zwick A, Bazinet AL, Cummings MP et al (2013) A large-scale, higher-level, molecular phylogenetic study of the insect order Lepidoptera (moths and butterflies). PLoS ONE. doi:10.1371/journal.pone.0058568

Roeder KD (1964) Aspects of the noctuid tympanic nerve response having significance in the avoidance of bats. J Insect Physiol 10(4):529–546

Roeder KD (1967) Turning tendency of moths exposed to ultrasound while in stationary flight. J Insect Physiol 13(6):873–888

Roeder KD, Treat AE (1957) Ultrasonic reception by the tympanic organ of noctuid moths. J Exp Zool 134(1):127–157

Rothstein SI (1990) A model system for coevolution: Avian brood parasitism. Annu Rev Ecol Syst 21:481–508

Russo D, Jones G, Arlettaz R (2007) Echolocation and passive listening by foraging mouse-eared bats *Myotis myotis* and *M blythii*. J Exp Biol 210(1):166–176

Rydell J, Yalden DW (1997) The diets of two high-flying bats from Africa. J Zool 242:69–76

Rydell J, Jones G, Waters D (1995) Echolocating bats and hearing moths: who are the winners? Oikos, 419–424

Rydell J, Kaerma S, Hedelin H, Skals N (2003) Evasive response to ultrasound by the crepuscular butterfly *Manataria maculata*. Naturwissenschaften 90(2):80–83

Schnitzler H-U, Flieger E (1983) Detection of oscillating target movements by echolocation in the Greater horseshoe bat. J Comp Physiol 153(3):385–391

Schnitzler H-U, Kalko EKV (2001) Echolocation by insect-eating bats. Bioscience 51(7):557–569

Schoeman CM, Jacobs DS (2003) Support for the allotonic frequency hypothesis in an insectivorous bat community. Oecologia 134(1):154–162

Schoeman MC, Jacobs DS (2008) The relative influence of competition and prey defenses on the phenotypic structure of insectivorous bat ensembles in southern Africa. PLoS ONE 3(11): e3715

Schoeman MC, Jacobs DS (2011) The relative influence of competition and prey defences on the trophic structure of animalivorous bat ensembles. Oecologia 166(2):493–506

Schuller G (1984) Natural ultrasonic echoes from wing beating insects are encoded by collicular neurons in the CF-FM bat, *Rhinolophus ferrumequinum*. J Comp Physiol A: Neuroethol Sens Neural Behav Physiol 155(1):121–128

Simmons NB, Seymour KL, Habersetzer J, Gunnell GF (2008) Primitive early Eocene bat from Wyoming and the evolution of flight and echolocation. 451:818–822. doi:10.1038/nature06549

Spangler HG (1988) Moth hearing, defense, and communication. Annu Rev Entomol 33:59–81

Stoffberg S, Jacobs DS, Matthee CA (2011) The divergence of echolocation frequency in horseshoe bats: moth hearing, body size or habitat? J Mammal Evol 18(2):117–129

Strauß J, Stumpner A (2015) Selective forces on origin, adaptation and reduction of tympanal ears in insects. J Comp Physiol A: Neuroethol Sens Neural Behav Physiol 201(1):155–169

Surlykke A (1988) Interaction between echolocating bats and their prey. Animal sonar. Springer, Berlin, pp 551–66

Surlykke A, Filskov M (1997) Hearing in *Geometrid moths*. Naturwissenschaften 84:356–359

Surlykke A, Skals N, Rydell J, Svensson M (1998) Sonic hearing in a diurnal geometrid moth, *Archiearis parthenias*, temporally isolated from bats. Naturwissenschaften 85(1):36–37

Teeling EC (2009) Hear, hear: the convergent evolution of echolocation in bats? Trends Ecol Evol 24(7):351–354

Teeling EC, Springer MS, Madsen O, Bates P, O'Brien SJ, Murphy WJ (2005) A molecular phylogeny for bats illuminates biogeography and the fossil record. Science 307(5709):580–584

ter Hofstede, HM, Ratcliffe JM (2016) Evolutionary escalation: the bat–moth arms race. J Exp Biol 219:1589–1602 doi:10.1242/jeb.086686

Vaughan N (1997) The diets of British bats (Chiroptera). Mammal Rev 27(2):77–94

Von Der Emde G, Schnitzler H-U (1986) Fluttering target detection in hipposiderid bats. J Compar Physiol A: Neuroethol Sens Neural Behav Physiol 159(6):765–772

Waters DA (2003) Bats and moths: what is there left to learn? Physiol Entomol 28(4):237–250

Waters D, Jones G (1996) The peripheral auditory characteristics of noctuid moths: responses to the search-phase echolocation calls of bats. J Exp Biol 199(4):847–856

Woodsworth G, Bell G, Fenton M (1981) Observations of the echolocation, feeding behaviour, and habitat use of *Euderma maculatum* (Chiroptera: Vespertilionidae) in southcentral British Columbia. Can J Zool 59(6):1099–1102

Yack JE (2004) The structure and function of auditory chordotonal organs in insects. Microsc Res Tech 63(6):315–337

Yack JE, Fullard JH (2000) Ultrasonic hearing in nocturnal butterflies. Nature 403(6767):265–266

Yack JE, Kalko EK, Surlykke A (2007) Neuroethology of ultrasonic hearing in nocturnal butterflies (Hedyloidea). J Comp Physiol A 193(6):577–590

Yamamoto S, Sota T (2007) Phylogeny of the Geometridae and the evolution of winter moths inferred from a simultaneous analysis of mitochondrial and nuclear genes. Mol Phylogenet Evol 44(2):711–723

Zahiri R, Kitching IJ, Lafontaine JD, Mutanen M, Kaila L, Holloway JD, Wahlberg N (2011) A new molecular phylogeny offers hope for a stable family level classification of the Noctuoidea (Lepidoptera). Zool Scr 40:158–173

Chapter 6
Co-evolution Between Bats and Frogs?

Abstract Besides moths, the only other bat–prey interaction, with the potential for co-evolution, is that with frogs. Bats can eavesdrop on sounds inadvertently produced by frogs as they move through their habitat or they can eavesdrop on the mating calls emitted by male frogs to attract females. Bats using such passive acoustic detection have an additional field of increased sensitivity outside the range of their own echolocation calls which matches frequencies at which most frogs call. Most frogs use audition for intra-specific communication, but as yet there is no evidence that they can hear echolocation. Some frog species are able to detect bats visually, allowing them to deploy a defence strategy before they are detected—the frog may cease movement or stop calling, depriving the bat of an acoustic cue. Frog mating calls are therefore under at least two strong selection pressures—female mate choice and bat predation. This may pose an evolutionary trade-off, potentially indicative of co-evolution, because call cessation has a negative effect on reproduction opportunities. However, although frogs appear to have evolved anti-bat behaviour, there is no indication that bats have responded. For example, some frogs have evolved calls that are difficult to localize for bats despite these calls being less attractive to females. The fringe-lipped bat may have responded by using water ripples produced by the calling frog and which remain available for a short time even after the frog stops calling. However, besides the latter case, no evidence has yet emerged that indicates an iterative reciprocal process between bats and frogs.

6.1 Overview

Many bats eat other vertebrates (e.g. other bats, fish, and frogs) but in the context of predator–prey interactions, that between bats and frogs is the best studied. These bat species usually live in environments in which the availability of prey types fluctuates widely and eat frogs opportunistically when they are abundant during the rainy season when frogs mate (Buchler and Childs 1981; Tuttle et al. 1981; Fenton et al. 1993; Seamark and Bogdanowicz 2002; Page and Ryan 2005; Shetty and Sreepada 2013). Frog-eating bats do not predominantly use echolocation when

© The Author(s) 2016

D.S. Jacobs and A. Bastian, *Predator–Prey Interactions: Co-evolution between Bats and their Prey*, SpringerBriefs in Animal Sciences, DOI 10.1007/978-3-319-32492-0_6 detecting frogs; instead, they detect their prey passively by eavesdropping on the sounds generated by frogs as they go about their lives. From a predator's perspective, frogs provide many cues that indicate their presence. Frogs produce rustling sounds when moving through leaf litter, splashes when jumping into water, waves when their vocal sac vibrates or they emit mating and territorial calls. As expected, these conspicuous acoustic cues are used not only by bats to prey on frogs but also by other taxa such as opossums, toads, storks, and parasitic flies (Jaeger 1976; Ryan et al. 1981; Tuttle and Ryan 1981; Bernal et al. 2006; Igaune et al. 2008).

Gathering information by eavesdropping on acoustic signals in the environment is a common strategy in bats. Echolocating bats have probably one of the most sophisticated auditory systems amongst animals providing precise sonar images of their surroundings (Simmons and Stein 1980; Neuweiler 1989; Schnitzler et al. 2003). When echolocating, they use sounds actively emitted by themselves to obtain detailed information encoded in the returning echo, but they also use their hearing to passively listen to sounds generated by other animals. These sounds can be emitted as social calls directed to the listener in a communication context, or they could be acoustic cues which were not specifically addressed to the listener, but which are nonetheless heard by the eavesdropping listener to obtain information. Bats, for example, can eavesdrop on the echolocation calls and the echoes therefrom to find new roosts (Barclay 1982; Ruczyński et al. 2009), feeding sites (Barclay 1982; Balcombe and Fenton 1988; Gillam 2007), to detect objects when flying in tandem with a conspecific (Koselj and Siemers 2013), to assess the species membership (Schuchmann and Siemers 2010; Bastian and Jacobs 2015) or gender (Schuchmann et al. 2012; Puechmaille et al. 2014), and of course to listen to the sounds of their prey such as the clicks from copulating flies (Siemers et al. 2012), mating calls of their insect (Chap. 4) and vertebrate prey (e.g. frogs).

Bat predation on frogs can vary in intensity from six small frogs per hour taken by the fringe-lipped bat, *Trachops cirrhosus*, (Ryan et al. 1981) to four medium-sized frogs a night eaten by *Nycteris grandis*, another frog-eating bat (Fenton et al. 1987). These predation rates by bats may represent a substantial selection pressure promoting the evolution of anti-predator strategies in frogs, which in turn could initiate the evolution of countermeasures to these strategies by bats. However, whether bat–frog interactions represent a co-evolved system is not clear. Here, we evaluate whether traits in bats and frogs have evolved as reciprocal and specific responses (Chap. 1) to traits in the other.

6.2   Audition in Bats and Frogs

Bats can eavesdrop on two different sound sources to detect and locate frogs (Fenton 1990); sounds inadvertently produced by frogs as they move through their habitat and mating calls emitted by male frogs to attract mates. The main difference between the two sounds is the origin of the sound and the reason the sound is generated in the first place. The rustling sound of a hopping frog in leaf litter, for example, is not actively produced, thus unintended by the frog and simply a by-product of its movement. Advertisement calls of frogs, on the other hand, are actively produced and emitted to attract females and to compete with other males (Ryan 1983; Gerhardt 1994b; Page et al. 2013). The acoustic conspicuousness of rustling sounds by hopping frogs depends on and fluctuates with the substrate surface and reveals little detail about the moving prey except maybe its size, direction of movement and distance from the bat. However, the vocalizations of frogs have evolved as species indicators which need to be reliably perceived and easily detectable by conspecifics. The information content of such signals is much higher and may allow a predatory bat to determine not only the location of the signaller but also its identity. Interestingly, frog-eating bat species do not make use of both sound sources to the same extent. Some species of bats exclusively use movement sounds as acoustic cues and ignore frog calls, whereas other bats primarily use frog vocalizations to find their prey. Representatives of bats using each of these strategies are amongst the best studied examples of frog-eating bats. The Indian false vampire bat, *Megaderma lyra*, attacks frogs only when they move and rustle and ignores them when they are stationary although vocalizing (Marimuthu and Neuweiler 1987). In contrast, the fringe-lipped bat, *T. cirrhosus*, homes in on the frogs' advertisement calls to find them. Most studies on *M. lyra* are of an experimental nature, focussing on the acoustic cues of the prey which are necessary to elicit an attack by the bat (Marimuthu and Neuweiler 1987; Marimuthu et al. 2002). Little to nothing is known about how *M. lyra* hunts frogs in the wild (Ratcliffe et al. 2005) and how the frogs respond to this predation.

   T. cirrhosus is by far the best studied frog-eating bat. Decades of integrative research provides a comprehensive understanding of various natural and sexual selection pressures, the role of phylogenetic signal as well as ultimate and proximate explanations of the evolution of the fringe-lipped bat, its frog prey and their potential co-evolution (Page et al. 2013). Information on the other frog-eating bat species is scarce but few of these species may provide alternative models for studying predator–prey dynamics between bats and frogs. For example, similar to *M. lyra*, *N. grandis* and *Cardioderma cor* use rustling sounds of moving insects and frogs but not the calls of frogs to locate and pursue them (Vaughan 1976; Fenton et al. 1983; Ryan and Tuttle 1987). So far, *T. cirrhosus* is the only known bat species exploiting the calls of frogs for foraging.

6.3 Predator: Detection of Frogs by Bats

Bats using passive acoustic detection exhibit sensory sensitivities for sounds of low frequency (<20 kHz) generated by the prey in addition to the sensory sensitivity of high frequency sounds of their own echolocation calls (Poussin and Simmons 1982; Ryan et al. 1983; Rübsamen et al. 1988; Bruns et al. 1989; Marimuthu 1997). Higher sensitivity means that the auditory system has lower thresholds for frequencies covering the range of rustling sounds as, for example, the low frequency hearing below 1 kHz in the case of *M. lyra* (Neuweiler et al. 1984; Marimuthu et al. 2002) or lower thresholds for those frequencies in which the frogs call as in the case of *Eptesicus fuscus* (Buchler and Childs 1981) or *T. cirrhosus* (Ryan and Tuttle 1983). The morphology of the ear of *T. cirrhosus* as well as its behavioural audiogram shows an additional field of increased sensitivity outside the range of their own echolocation calls which matches the frequencies of below 5 kHz at which most frogs call including the tungara frog, *Engystomops* (=*Physalaemus*) *pustulosus* (Ryan et al. 1983; Bruns et al. 1989). Although there are several studies on frog-eating bats and their mode of foraging, most knowledge on the interaction between bats and frogs and specifically on the frogs' role is drawn from the work on *T. cirrhosus* and *E. pustulosus* (for a review see Ryan 2011).

6.4 Prey: The Frogs' Adaptive Responses to the Detection by Bats

6.4.1 Predator Avoidance Abilities (Pre-detection Primary Defence)

Perhaps the most effective anti-predator defence is to avoid being detected by the predator. Once detected, prey have to either escape through fleeing or avoid its position being located by the predator by suppressing the cues that allowed detection in the first place. Both, active advertisement calls and passive movement sounds are inevitably provided cues to eavesdroppers by the frogs. However, the options for potentially adaptive anti-predator responses by the prey may differ with respect to these two kinds of acoustic cues. With respect to rustling sounds, the anti-predator behaviour that prevents bats detecting the frog may be limited because the frog cannot change the acoustic properties of the surface it is moving on. As an adaptive response to bat predation, frogs may avoid moving on 'loud' surfaces but we are not aware of any observations to this effect. Of course, if a frog becomes aware of the bat before the bat detects the frog, the frog may cease movement depriving the bat of an acoustic cue. Again we are not aware of any research done on the acoustic startle response of frogs in this context.

Most frog species emit advertisement calls (Gerhardt and Huber 2002; Ryan 2005) and as advertisement calls are actively emitted by frogs and are used by predatory bats to attack them, potential adaptive co-evolution dynamics between predator and prey are plausible. Advertisement calls of male frogs are used to attract females which use these calls to select their mates (Gerhardt 1994b; Gerhardt and Huber 2002). Due to its central part within the reproductive behaviour of frogs, calling is essential for male frogs. Females prefer males with acoustically more elaborated variants of calls with higher background contrast (Bernal et al. 2009; Baugh and Ryan 2010; Page et al. 2013) and so does the predatory bat *T. cirrhosus*

(Ryan et al. 1982). In some frog species, e.g. *Leptodactylus* species from Brazil (Prado and Haddad 2003), non-calling males without territory, called satellite males, intercept females which were attracted to the actual calling male. The female frogs mistake the satellite male as the caller (Gerhardt and Huber 2002; Wells 2007). Satellite males are less conspicuous to bats because they do not call. In *E. pustulosus*, such a strategy was never observed (Ryan 1983). Instead, males reduce the risk of predation by not attending the chorus on every night during the mating season. This strategy was attributed to the high predation risk when calling (Green 1990). How these males compensate for the loss of reproductive opportunities as a result of not attending the chorus is not known. Thus, the evolution of a strategy which leads to avoiding predation by simply not producing the conspicuous cue carries the risk of reduced mating and reproduction in the species in which female choice is based on the male advertisement calls (Green 1990; Gerhardt 1994a; Lahanas 1995; Gerhardt and Huber 2002; Wells 2007). Alternatively, a potential adaptive response to predation may manifest itself in the form of a shift of the signals' acoustic properties away from the bats sensitivity field to avoid detection (Zuk and Kolluru 1998), an adaptation for private communication channels as found in other predator–prey systems (Zuk and Kolluru 1998; Stoddard 1999; Cummings et al. 2003; Römer et al. 2010; Caldart et al. 2014). There are, however, some circumstances which make such signal divergence less likely. For one, the bats auditory sensitivity could follow this shift in a reciprocal manner in the same way it initially evolved additional hearing sensitivities for prey generated sounds. The low-frequency hearing in *T. cirrhosus* has likely evolved to improve the perception of the tungara frog calls' (Ryan et al. 1983; Bruns et al. 1989) and the auditory system may therefore have the necessary plasticity to follow acoustic changes in the frogs' calls.

Moreover, competing selection forces, which may favour more elaborate signals, affect signal design as well. The advertisement call of frogs is under female selection (Ryan 1986). For *E. pustulosus*, it was shown that sensory exploitation of female preferences likely played a role in the evolution of the males call (but see Ron 2008; Jones et al. 2014). Male call attributes such as lower frequencies evolved to exploit pre-existing preferences in females (Ryan and Rand 2003; Ryan 2005) probably because they elicit increased sensory stimulation (Seeba et al. 2010) as lower frequencies match better with the sensitivity centre of the basilar papilla (Ryan 2005), one of the two hearing organs in frogs. In addition, both vocally competing males and choosing females (Tuttle and Ryan 1982) respond more strongly to more complex acoustic signals (Ryan 2011). Furthermore, sexual selection for advertisement signals, such as mating calls, must be linked with female preference for those signals so that they can evolve in tandem (Fisher 1930). Shifts in the advertisement calls therefore require shifts in the auditory pathway and advertisement behaviour of both males and females. Besides sexual selection, natural selection exerted by habitat characteristics also influences sound propagation of the signal (Wiley and Richards 1978, Wells and Schwartz 1982; Endler 1992; Forrest 1994; Barth and Schmid 2001). The call must travel far enough through the habitat to reach females and neighbouring males. Any signal divergence which affects, e.g. the range of the signal carries the risk of it not being effectively propagated through the habitat. This suite of selection pressures may limit or even oppose changes to the acoustic characteristics of the call that make it more inconspicuous to predatory bats.

Alternatively to changes in the signal itself, a 'response' of frogs to bat predation might be the synchronization of calling (Tuttle and Ryan 1982). For example, in the Panama cross-banded treefrog, *Smilisca sila*, males have less risk of predation by bats when they synchronize their calls with other males. The mechanisms of localization of frog calls by bats involve minute fluctuations in binaural and monaural information within the returning echoes (Popper and Fay 1995; Schnitzler and Kalko 2001; Popper and Fay 2005). In synchronized calls, bats probably experience the difficulties in localizing an individual bat due to binaural cross-correlations (Tuttle and Ryan 1982; Popper and Fay 2005). When experimentally tested, *T. cirrhosus* responded more often to the asynchronous than towards the synchronized calls of tree frogs by flying towards the loudspeaker through which the calls were broadcast (Tuttle and Ryan 1982). Predator confusion due to more complex acoustic environments can be an effective strategy but a simple dilution effect of big groups (Davies et al. 2012; Rhebergen et al. 2015) also lowers the predation risk for an individual frog (Ryan et al. 1981). The trade-off in this scenario for the male frog is the increased number of rivals competing for females (Ryan et al. 1981; Benedix and Narins 1999). This negative effect may be counterbalanced by the improved sex ratio at the pond because the bigger the chorus the more females that are attracted (Ryan et al. 1981; Brush and Narins 1989).

6.4.2 Capture Avoidance Strategies (Post-detection Secondary Defence)

6.4.2.1 Detection of Predatory Bats by Frogs

In cases where frogs cannot avoid being detected by bats, the next line of defence would be to avoid being captured by an attacking bat. Most bats use acoustic information to pinpoint the exact position of the prey before capture (Popper and Fay 1995; Schnitzler and Kalko 2001). Fleeing silently or the immediate suppression of the acoustic cue that bats use to localize the frog may therefore be effective strategies by frogs to escape predation. However, the frogs will have to be able to detect the foraging bat to employ one or the other of these escape strategies. This can be done through audition. Acoustic detection of bats has evolved several times in tympanate moths which have auditory sensors sensitive to the echolocation frequencies of their respective predators (Chap. 4). Frogs communicate acoustically, so they are equipped with an auditory sense receptor in form of a tympanum (Wells 2007). In addition, a few frog species emit ultrasonic calls to communicate

(Narins et al. 2004; Feng et al. 2006; Gridi-Papp et al. 2008; Arch et al. 2009; Shen et al. 2011), most likely to avoid masking by intense low-frequency noise of their habitat. This sensitivity to ultrasound may also allow them to detect the ultrasonic echolocation calls of bats.

Most frog species, however, communicate in lower frequency ranges and might therefore not be able to detect the echolocation calls (Loftus-Hills and Johnstone 1970; Gerhardt and Schwartz 2001). So far, no study has reported acoustic detection of bat echolocation calls by frogs. We cannot report on any reciprocal adaptation of the bats in this regard because there is as yet no indication that frogs can hear and respond to the echolocation calls of attacking bats.

Frogs may also detect an approaching bat visually, in which case the visual receptor must be sensitive enough to detect movements and/or shapes of bats under dimmed light conditions early enough to allow the frog to respond. Frogs also use visual signals to communicate, but thus far it is only *S. sila* and *E. pustulosus* that are known to use vision to detect an approaching bat provided that sufficient moonlight is present (Tuttle and Ryan 1982; Tuttle et al. 1982; Bulbert et al. 2015). *E. pustulosus* can discriminate the potentially dangerous predatory bats from non-dangerous bats on the basis of its shape when flying (Tuttle et al. 1982).

6.4.2.2 Avoidance of Capture

Frogs use a variety of secondary defences to avoid capture or being eaten once they have been detected by a bat. Some frog species secrete unpalatable or even poisonous substances from their skin in response to predation pressure (Wells 2007). Frogs, in which this defence strategy has not evolved can escape an attacking bat that detected them by fleeing. The frogs may dive under water after detecting a foraging bat (Tuttle et al. 1982; Bernal et al. 2007b) or undercut the flight path of the attacking bat (Bulbert et al. 2015). Fleeing to a location inaccessible for the bat can be successful but any movement may again generate rustling sounds and water movements and thus provide cues for pursuing bats (Suthers 1965; Siemers et al. 2001; Halfwerk et al. 2014b).

The frog may detect the approaching bat and suppress rustling sounds or calling, and remaining motionless until the danger is over (Bulbert et al. 2015). Bats rely heavily on acoustic cues when detecting, classifying and locating prey (Neuweiler 1990; Schnitzler and Kalko 2001). For passive locators such as the frog-eating bats, these acoustic cues are not self-generated as is echolocation. These bats depend on the continuous generation of sound by the prey during the pursuit. If the prey suppresses the signal, the bat loses track of it, reducing the bat's capture success (Bulbert et al. 2015). Experiments have shown that *T. cirrhosus* capture attempts on frogs become less accurate when the acoustic cue suddenly ceases (Page and Ryan 2008).

However, suppression of calling also directly impacts on the reproductive success of males because females base their mate choice on a comparison of the calls of several males (Page et al. 2013). The females' working memory of calls emitted prior to call cessation is limited and requires male calls to be emitted at more or less the same time for comparison (Akre and Ryan 2010).

Adaptive responses affecting the emission of advertisement calls are under restrictions due to their direct importance for reproductive success. Male frogs cannot 'whisper' (Johnstone 1998) because females might be far and not hear the softer calls. The calls also need to be elaborate to compete effectively against the calls from other males and to advertise attractiveness to females (Ryan 2011). Thus, over evolutionary time female choice has resulted in male calls becoming increasingly elaborate (Johnstone 1998; Trillo et al. 2013).

In *E. pustulosus*, for example, vocal competition in larger choruses increased the rate of calling and the number of complex calls emitted (a more elaborate signal) which attract more females and may also increase the chances of obtaining a mate (Green 1990). Complex calls consist of two call components, the 'whine' and the 'chuck', whereas a simple call consists only of 'whines'. However, more elaborate calls of course attract not only females but also the bat *T. cirrhosus* which also displayed a preference for complex calls (Akre et al. 2011; Trillo et al. 2013) and higher call rates (Ryan et al. 1982). Such preference in bats is apparently an adaptation resulting from an association between call complexity and higher prey densities (Bernal et al. 2007a; Fugere et al. 2015) but also allows increased localization acuity which equates to increased capture success (Page and Ryan 2008).

Yet another strategy to reduce the risk of being captured is to call less, especially when the detection of hunting bats in time is not secured. The treefrog *S. sila* calls less and the calls are less conspicuous (i.e. less complex and no 'chucks' added) on nights with little moonlight when bats are difficult to spot (Nunes 1988). Similarly, *E. pustulosus* uses less complex calls and generally calls less often on dark nights when bats are difficult to spot (Tuttle and Ryan 1982). Strong selection against complex calls in response to predation was also found in a study comparing signal divergence across a geographical range differing in bat predation pressure (Trillo et al. 2013). When a flying *T. cirrhosus* was visually detected by *E. pustulosus*, the entire chorus rapidly responds with cessation of calling within a second (Tuttle et al. 1982; Dapper et al. 2011). The speed at which the frog chorus shuts down is correlated with the urgency of the alarm in the form of proximity (flight height) of the predator (Tuttle et al. 1982; Lima and Dill 1990). At least in *E. pustulosus*, cessation of calling is a selective response towards the predation by *T. cirrhosus* because they remained silent for about eight times longer than after a normal, random chorus shutdown or after detection of non-predatory bats (Tuttle et al. 1982). Reducing the calling time has a negative effect on mating success (Baugh and Ryan 2010), because females actively choose from amongst calling males and males compete acoustically with neighbouring males. Another strategy to avoid eavesdropping bats is calling from more concealed places in the presence of the predatory bat *T. cirrhosus* (Delia et al. 2010) or on nights with less moon light (Höbel 1999).

6.5 Predator: The Bats' Response to Anti-bat Strategies

Bats responded to tympanate moths by emitting very faint, 'whispering' echolocation calls (Chap. 4), but because frogs do not use the echolocation calls of bats to detect and avoid them, different counter adaptations to prey strategies may evolve.

6.5.1 Adaptive Responses to the Frogs' Anti-bat Strategies

Frog-eating bat species such as *T. cirrhosus*, *M. lyra*, *M. spasma* and *C. cor* possess salivary glands, which may play a role in dealing with unpalatable secretions of frogs. The accessory submandibular glands of these species deviate histologically from the typical mammalian submandibular gland. Presumably, these deviations reflect differences in function, possibly involving the secretion of factors protecting the bat from toxins in the integument of some frog species (Tandler et al. 1997). However, experimental observations suggest otherwise (see below). These glands supposedly evolved independently in *T. cirrhosus* and other species, apparently in response to their frog-eating habits (Phillips et al. 1987; Tandler et al. 1996, 1997). In experiments with *M. lyra* and *T. cirrhosus* in which a naturally palatable frog was covered with an unpalatable frogs' or toads' secretions, bats captured the frog but then detected the substance and dropped the frog (Marimuthu and Neuweiler 1987; Page et al. 2012). Furthermore, *T. cirrhosus* has never been observed in the wild attempting to capture unpalatable frog species (Tuttle and Ryan 1981). The bats use the calls of the different frog species to discriminate between palatable and unpalatable frogs (Ryan and Tuttle 1983; Page and Ryan 2006).This is at least in part a learned behaviour (Page and Ryan 2005, 2006; Jones et al. 2013b) which would explain the evolution of accessory salivary glands. This adaptation does not 'overcome' the frogs anti-predator strategy of producing poisonous secretions as an arms race would require but rather shows that bats adapted to poisonous frogs by detecting the toxins and avoiding these frogs, breaking the interaction that may have led to an iterative process of co-evolution.

Bats that glean using prey generated sounds to detect prey on the substrate will in most cases catch the prey directly from the substrate and not during flight (Fenton 1990; Denzinger and Schnitzler 2013). Prey that call from concealed locations, e.g. under leafs might prevent gleaning bats from successfully detecting them (Höbel 1999). This strategy may not however work against the New Zealand bat species, *Mystacina tuberculata*, which crawls through leaf litter in the search of prey (Jones et al. 2003). The strategy to call from concealed places is assumed to be the reason why other gleaning bats do not use the calls of crickets to detect them but rely on the sounds cricket generate when moving around (Fuzessery et al. 1993).

Visual detection of bats by frogs becomes increasingly difficult with increasing darkness which means in turn that bats hunting in moonlit nights have reduced chances to catch their prey (Tuttle et al. 1982; Lima and Dill 1990; Dapper et al. 2011). Indeed, the numbers of bats emerging on full-moon nights drop noticeably but this can also be correlated with increased predation risk for the bats themselves by e.g. owls during full moon (Usman et al. 1980; Lang et al. 2006; Lima and O'Keefe 2013; Thomas and Jacobs 2013).

6.5.2 Novel Predator Abilities Enhancing Prey Detection and Capture

In a co-evolving system, it might be expected that bats respond by evolving hunting strategies that overcome the prey's defences and improve the bats' success rate. For example bats could increasing the speed (Jones et al. 2013a) or the angle (Rhebergen et al. 2015) of attack, thus reducing the time the frog has to respond after it has detected the danger, or switch to another sensory modality (e.g. echolocation).

As mentioned before, *T. cirrhosus* fails in its capture attempts if the guiding cue, namely the frogs' call, ceases (Page and Ryan 2008). However, bats can make use of echolocation to detect movements of the vocal sac of calling frogs (Halfwerk et al. 2014a) and in addition to that use the water ripples produced by the vocal sac movement to localize the frog (Halfwerk et al. 2014b). These water ripples remain detectable for up to 3 s after the frog ceased calling (Halfwerk et al. 2014b) and provide longer-lasting cues for the bat to locate the frog and therefore might reflect an adaptive response by the bat to call cessation.

6.6 Conclusion

The predator–prey dynamics between bats and frogs, mainly between *T. cirrhosus* and *E. pustulosus*, contain some reciprocity in the form of adaptive responses to each other's traits associated with predation. For example, the auditory sense of *T. cirrhosus* has evolved additional sensitivities to frog calls and the frogs may have reciprocally evolved the ability to detect approaching bats by their shape and cease calling. However, besides reciprocity, specificity is a crucial component of an arms race. The bat–frog systems so far studied do not show any evidence of specificity (Sect. 1.2).

Similarly, the co-evolution between frogs and those bats which use motion generated sounds holds little potential for the initiation of an arms race. This is because the rustling sound made by the frog as it moves through the vegetation is largely not under the control of the frog and there is no evidence that frogs have evolved any strategies to reduce the conspicuousness of such sounds in direct response to bat predation. Natural selection acting on mating calls instead of rustling sounds in contrast provides several opportunities for the initiation of an arms race, at least hypothetically. It is plausible that frogs could respond directly to eavesdropping by bats by evolving calls that are more difficult for the bat to detect. This would satisfy both the reciprocity and specificity criteria of co-evolution, especially if bats in turn evolved greater hearing acuity to detect frog calls. The bat *T. cirrhosus* has evolved additional hearing sensitivity to hear frog advertisement calls but there is no evidence that frogs have responded by evolving less conspicuous calls. As shown for *E. pustulosus*, the frogs' advertisement call has little leeway to diverge due to strong and opposing selection pressures on call characteristics. Sexual selection by female mate choice favours more conspicuous calls and sound propagation characteristics of the habitat favour calls which can spread far through the habitat. This is contrary to the less conspicuous calls needed to avoid bats. Selection pressures exerted by the habitat and female choice thus lead to elaborated signals which are easier and reliably received by female and male frogs but also by bat predators.

Alternatively, instead of evolving less conspicuous calls, frogs could avoid the disadvantages of such calls by evolving the ability to detect the echolocation calls of bats before the bats detect the frogs. Echolocation can alert the prey of the predators' presence if the prey has ultrasonic hearing, i.e. hearing above 20 kHz (Speakman 1993). Passive listening by bats for prey generated sounds is assumed to be a secondarily evolved function of its auditory system in response to ultrasonic hearing in prey (Speakman 1993; Sales 2012). Frog-eating bats do not predominantly rely on echolocation when detecting their prey, with the exception of *M. lyra* (Marimuthu et al. 1995; Schmidt et al. 2000), whereas moth-hunting bats detect their prey through echolocation (Neuweiler 1980). However, bats that use passive listening are not mute, and they emit echolocation calls during the attack (Barclay et al. 1981; Fenton et al. 1983; Ryan and Tuttle 1987; Fenton 1990; Fuzessery et al. 1993; Marimuthu et al. 1995; Schmidt et al. 2010) mostly but not solely to orientate and to determine distances to obstacle or the surface from which they glean their prey.

Besides *T. cirrhosus*, frog-eating bats are not specialized predators of frogs. They glean ground-dwelling insects as well as different vertebrates, including other amphibians, fish and small mammals, and some of these prey taxa might be able to detect and respond to echolocation calls. Generalists such as most of the frog-eating bat species may employ both foraging modes, active echolocation and passive audition, depending on the prey type they pursue (Bonato and Facure 2000). This plasticity in foraging mode is supported by experimental work on *M. lyra*, in which the bats used faint echolocation to detect motionless frogs in water where it had previously experienced the presence of frogs, but relied on passive audition and no echolocation when frogs where actively moving on the substrate (Marimuthu et al. 1995). In another experiment, *M. lyra* also detected and captured motionless mice or mealworms in the absence of passive acoustic cues solely by echolocation (Schmidt et al. 2000). Similarly, *T. cirrhosus* makes use of echolocation when assessing prey quality such as palatability of frogs and when hunting in cluttered habitats (Page et al. 2012; Surlykke et al. 2013). Thus, echolocation calls are available as cues directly linked to the predator which could be used frogs to detect bats. This could lead to an adaptive response in frogs and ultimately increasing the frogs' probability of escape. As in moths, this would in turn exert selection pressure on the echolocation emission by bats and so on. However, this evolutionary process has not yet been found in bat–frog systems (Norberg and Fenton 1988).

One explanation why co-evolution has not occurred might be the characteristics of the auditory system of frogs. Frogs have two inner ear organs, the amphibian papilla (AP) sensitive to lower frequencies and the basilar papilla (BP) sensitive to higher frequencies (Wells 2007). According to the Capranicas matched filter hypothesis and studies supporting it, both centres of sensitivity are tuned with the frequency peaks of species-specific calls probably to improve the detection of conspecific signals against the background noise of the environment (Capranica and Moffat 1983; Gerhardt and Schwartz 2001). In *E. pustulosus* these two hearing centres are somewhat matched to the two call components that make up a complex advertisement call (Ryan 1990), the 'whine' (700 Hz) which identifies the species and mate (assortative mating) and the 'chuck' (2500 Hz) which plays a role in male competition and female preference (selective mating) (Ryan and Rand 2003; Ryan 2005). Most frog calls have frequencies well below echolocation calls of bats (Loftus-Hills and Johnstone 1970; Feng et al. 2006) and the detection of ultrasound requires an additional sensitivity centre and respective behavioural, neurophysio-logical and anatomical adaptations. The frog's audition system may thus be con-strained from evolving anti-bat systems based on audition.

There is little or no detailed knowledge on any other bat–frog system besides *T. cirrhosus* and *E. pustulosus*, especially from the frogs' perspective. Focussing research on the frogs' perspective may therefore offer additional insights. As cor-rectly pointed out by Bernal et al. (2007b), little research has been done on acoustic predator recognition (Schwartz et al. 2000; Llusia et al. 2010) despite the impor-tance of the acoustic sense in frogs (Gerhardt and Huber 2002). In this context, it would be of great interest to investigate if ultrasound hearing in anurans, in general, as described in recent research (Narins et al. 2004; Feng et al. 2006; Gridi-Papp et al. 2008; Arch et al. 2009; Shen et al. 2011) is present in frog species which are preyed upon by bats and if they can acoustically detect bats. Such a scenario would provide a potential candidate for another arms race in which bats are involved.

References

Akre KL, Ryan MJ (2010) Complexity increases working memory for mating signals. Curr Biol 20(6):502–505
Akre KL, Farris HE, Lea AM, Page RA, Ryan MJ (2011) Signal perception in frogs and bats and the evolution of mating signals. Science 333(6043):751–752
Arch VS, Grafe TU, Gridi-Papp M, Narins PM (2009) Pure ultrasonic communication in an endemic Bornean frog. PLoS ONE 4(4):e5413
Balcombe J, Fenton MB (1988) Eavesdropping by bats: the influence of echolocation call design and foraging strategy. Ethology 79(2):158–166

Barclay RMR (1982) Interindividual use of echolocation calls: eavesdropping by bats. Behav Ecol Sociobiol 10(4):271–275

Barclay RMR, Fenton M, Tuttle M, Ryan M (1981) Echolocation calls produced by *Trachops cirrhosus* (Chiroptera: Phyllostomatidae) while hunting for frogs. Can J Zool 59(5):750–753

Barth FG, Schmid A (2001) Ecology of sensing (1 ed). Springer, Berlin

Bastian A, Jacobs DS (2015) Listening carefully: increased perceptual acuity for species discrimination in multispecies signalling assemblages. Anim Behav 101:141–154

Baugh AT, Ryan MJ (2010) Mate choice in response to dynamic presentation of male advertisement signals in tungara frogs. Anim Behav 79(1):145–152

Benedix J Jr, Narins PM (1999) Competitive calling behavior by male treefrogs, *Eleutherodactylus coqui* (Anura: Leptodactylidae). Copeia 1118–1122

Bernal XE, Rand AS, Ryan MJ (2006) Acoustic preferences and localization performance of blood-sucking flies (*Corethrella Coquillett*) to tungara frog calls. Behav Ecol 17(5):709–715

Bernal XE, Stanley Rand A, Ryan MJ (2007a) Sexual differences in the behavioral response of tungara frogs, *Physalaemus pustulosus*, to cues associated with increased predation risk. Ethology 113(8):755–763

Bernal XE, Page RA, Rand AS, Ryan MJ (2007b) Cues for eavesdroppers: do frog calls indicate prey density and quality? Am Nat 169(3):409–415

Bernal XE, Akre KL, Baugh AT, Rand AS, Ryan MJ (2009) Female and male behavioral response to advertisement calls of graded complexity in tungara frogs, Physalaemus pustulosus. Behav Ecol Sociobiol 63(9):1269–1279

Bonato V, Facure K (2000) Bat predation by the fringe-lipped bat *Trachops cirrhosus* (Phyllostomidae, Chiroptera). Mammalia-Paris 64(2):241–242

Bruns V, Burda H, Ryan MJ (1989) Ear morphology of the frog-eating bat (*Trachops cirrhosus*, family: Phyllostomidae): apparent specializations for low-frequency hearing. J Morphol 199 (1):103–118

Brush JS, Narins PM (1989) Chorus dynamics of a Neotropical amphibian assemblage: comparison of computer simulation and natural behaviour. Anim Behav 37:33–44

Buchler E, Childs S (1981) Orientation to distant sounds by foraging big brown bats (*Eptesicus fuscus*). Anim Behav 29(2):428–432

Bulbert MW, Page RA, Bernal XE (2015) Danger comes from all fronts: Predator-dependent escape tactics of tungara frogs. PLoS ONE 10(4):12

Caldart VM, Iop S, Cechin SZ (2014) Social interactions in a neotropical stream frog reveal a complex repertoire of visual signals and the use of multimodal communication. Behaviour 151 (6):719–739

Capranica RR, Moffat AJ (1983) Neurobehavioral correlates of sound communication in anurans. In: Moffat AJ, Capranica RR, Ingle DJ (eds) Advances in vertebrate neuroethology, vol 56. Springer, New York, pp 701–730

Cummings ME, Rosenthal GG, Ryan MJ (2003) A private ultraviolet channel in visual communication. Proc Royal Soc London B: Biol Sci 270(1518):897–904

Dapper AL, Baugh AT, Ryan MJ (2011) The sounds of silence as an alarm cue in tungara frogs, Physalaemus pustulosus. Biotropica 43(3):380–385

Davies NB, Krebs JR, West SA (2012) An introduction to behavioural ecology. Wiley, Oxford, UK

Delia J, Cisneros-Heredia DF, Whitney J, Murrieta-Galindo R (2010) Observations on the reproductive behavior of a neotropical Glassfrog, *Hyalinobatrachium fleischmanni* (Anura: Centrolenidae). S Am J Herpetol 5(1):1–12

Denzinger A, Schnitzler H-U (2013) Bat guilds, a concept to classify the highly diverse foraging and echolocation behaviors of microchiropteran bats. Front Physiol 4:1–15

Endler JA (1992) Signals, signal conditions, and the direction of evolution. Am Nat 139: S125–S153

Feng AS, Narins PM, Xu CH, Lin WY, Yu ZL, Qiu Q, Xu ZM, Shen JX (2006) Ultrasonic communication in frogs. Nature 440(7082):333–336

Fenton MB (1990) The foraging behavior and ecology of animal-eating bats. Can J Zool-Revue Canadienne De Zoologie 68(3):411–422

Fenton MB, Gaudet CL, Leonard ML (1983) Feeding-behavior of the bats *Nycteris grandis* and *Nycteris thebaica* (Nycteridae) in captivity. J Zool 200:347–354

Fenton MB, Cumming D, Hutton J, Swanepoel C (1987) Foraging and habitat use by *Nycteris grandis* (Chiroptera: Nycteridae) in Zimbabwe. J Zool 211(4):709–716

Fenton MB, Rautenbach IL, Chipese D, Cumming MB, Musgrave MK, Taylor JS, Volpers T (1993) Variation in foraging behavior, habitat use, and diet of Large slit-faced bats (*Nycteris grandis*). Zeitschrift für Säugetierkunde Biol 58(2):65–74

Fisher RA (1930) The genetical theory of natural selection: a complete, variorum edn. Oxford University Press. Edition, Oxford

Forrest T (1994) From sender to receiver: propagation and environmental effects on acoustic signals. Am Zool 34(6):644–654

Fugere V, Teague O'Mara M, Page RA (2015) Perceptual bias does not explain preference for prey call adornment in the frog-eating bat. Behav Ecol Sociobiol 69(8):1353–1364

Fuzessery ZM, Buttenhoff P, Andrews B, Kennedy JM (1993) Passive sound localization of prey by the pallid bat (*Antrozous p. pallidus*). J Comp Physiol A: Neuroethol Sens Neural Behav Physiol 171(6):767–777

Gerhardt HC (1994a) Reproductive character displacement of female mate choice in the grey treefrog, Hyla chrysoscelis. Anim Behav 47(4):959–969

Gerhardt CH (1994b) The evolution of vocalization in frogs and toads. Annu Rev Ecol Syst 25:293–324

Gerhardt HC, Huber F (2002) Acoustic communication in insects and anurans: common problems and diverse solutions. The University of Chicago Press, Chicago

Gerhardt HC, Schwartz JJ (2001) Auditory tuning and frequency preferences in anurans. Anuran Commun 73–85

Gillam E (2007) Eavesdropping by bats on the feeding buzzes of conspecifics. Can J Zool 85 (7):795–801

Green AJ (1990) Determinants of chorus participation and the effects of size, weight and competition on advertisement calling in the tungara frog, *Physalaemus pustulosus* (Leptodactylidae). Anim Behav 39(4):620–638

Gridi-Papp M, Feng AS, Shen J-X, Yu Z-L, Rosowski JJ, Narins PM (2008) Active control of ultrasonic hearing in frogs. Proc Natl Acad Sci 105(31):11014–11019

Halfwerk W, Jones PL, Taylor RC, Ryan MJ, Page RA (2014a) Risky ripples allow bats and frogs to eavesdrop on a multisensory sexual display. Science 343(6169):413–416

Halfwerk W, Dixon MM, Ottens KJ, Taylor RC, Ryan MJ, Page RA, Jones PL (2014b) Risks of multimodal signaling: bat predators attend to dynamic motion in frog sexual displays. J Exp Biol 217(17):3038–3044

Höbel G (1999) Notes on the natural history and habitat use of *Eleutherodactylus fitzingeri* (Anura: Leptodactylidae). Amphibia-Reptilia 20(1):65–72

Igaune K, Krams I, Krama T, Bobkova J (2008) White storks *Ciconia ciconia* eavesdrop on mating calls of moor frogs Rana arvalis. J Avian Biol 39(2):229–232

Jaeger RG (1976) A possible prey-call window in anuran auditory perception. Copeia 1976 (4):833–834

Johnstone RA (1998) Conspiratorial whispers and conspicuous displays: games of signal detection. Evolution 1554–1563

Jones G, Webb PI, Sedgeley JA, O'Donnell CF (2003) Mysterious Mystacina: how the New Zealand short-tailed bat (*Mystacina tuberculata*) locates insect prey. J Exp Biol 206(Pt 23):4209–4216

Jones PL, Ryan MJ, Flores V, Page RA (2013a) When to approach novel prey cues? Social learning strategies in frog-eating bats. Proc Royal Soc B—Biol Sci 280(1772):2013–2330

Jones PL, Farris HE, Ryan MJ, Page RA (2013b) Do frog-eating bats perceptually bind the complex components of frog calls? J Comp Physiol A: Neuroethol Sens Neural Behav Physiol 199(4):279–283

Jones PL, Ryan MJ, Page RA (2014) Population and seasonal variation in response to prey calls by an eavesdropping bat. Behav Ecol Sociobiol 68(4):605–615

Koselj K, Siemers BM (2013) Horseshoe bats can use information in echoes of conspecific calls for spatial orientation. Paper presented at the International Bat Research Conference, Costa Rica

Lahanas PN (1995) The function of near neighbors in decreasing call latency period by the tungara frog, Physalaemus pustulosus. Biotropica 27(2):262–265

Lang AB, Kalko EK, Römer H, Bockholdt C, Dechmann DK (2006) Activity levels of bats and katydids in relation to the lunar cycle. Oecologia 146(4):659–666

Lima SL, Dill LM (1990) Behavioral decisions made under the risk of predation: a review and prospectus. Can J Zool 68(4):619–640

Lima SL, O'Keefe JM (2013) Do predators influence the behaviour of bats? Biol Rev 88(3):626–644

Llusia D, Márquez R, Beltrán JF (2010) Non-selective and time-dependent behavioural responses of common toads (Bufo bufo) to predator acoustic cues. Ethology 116(12):1146–1154

Loftus-Hills JJ, Johnstone BM (1970) Auditory function, communication, and the brain-evoked response in anuran amphibians. J Acoust Soc Am 47(4B):1131–1138

Marimuthu G (1997) Stationary prey insures life and moving prey ensures death during the hunting flight of gleaning bats. Curr Sci 72(12):928–931

Marimuthu G, Neuweiler G (1987) The use of acoustical cues for prey detection by the Indian false vampire bat, Megaderma lyra. J Comp Physiol A: Neuroethol Sens Neural Behav Physiol 160(4):509–515

Marimuthu G, Habersetzer J, Leippert D (1995) Active acoustic gleaning from the water-surface by the Indian false vampire bat, Megaderma lyra. Ethology 99(1):61–74

Marimuthu G, Rajan KE, Kandula S, Parsons S, Jones G (2002) Effects of different surfaces on the perception of prey-generated noise by the Indian false vampire bat Megaderma lyra. Acta Chiropterologica 4(1):25–32

Narins PM, Feng AS, Lin WY, Schnitzler H-U, Denzinger A, Suthers RA, Xu CH (2004) Old World frog and bird, vocalizations contain prominent ultrasonic harmonics. J Acoust Soc Am 115(2):910–913

Neuweiler G (1980) How bats detect flying insects. Phys Today 33(8):34–40

Neuweiler G (1989) Foraging ecology and audition in echolocating bats. Trends Ecol Evol 4 (6):160–166

Neuweiler G (1990) Auditory adaptations for prey capture in echolocating bats. Physiol Rev 70 (3):615–641

Neuweiler G, Singh S, Sripathi K (1984) Audiograms of a South Indian bat community. J Comp Physiol A: Neuroethol Sens Neural Behav Physiol 154(1):133–142

Norberg UM, Fenton MB (1988) Carnivorous bats? Biol J Linn Soc 33(4):383–394

Nunes VD (1988) Vocalizations of treefrogs (Smilisca sila) in response to bat predation. Herpetologica 44(1):8–10

Page RA, Ryan MJ (2005) Flexibility in assessment of prey cues: frog-eating bats and frog calls. Proc Royal Soc Lond B: Biol Sci 272(1565):841–847

Page RA, Ryan MJ (2006) Social transmission of novel foraging behavior in bats: frog calls and their referents. Curr Biol 16(12):1201–1205

Page RA, Ryan MJ (2008) The effect of signal complexity on localization performance in bats that localize frog calls. Anim Behav 76:761–769

Page RA, Schnelle T, Kalko EKV, Bunge T, Bernal XE (2012) Sequential assessment of prey through the use of multiple sensory cues by an eavesdropping bat. Naturwissenschaften 99 (6):505–509

Page RA, Ryan MJ, Bernal XE (2013) Be loved, be prey, be eaten. In K Yasukawa (ed) Animal behavior (vol 3, Case studies: integration and application of animal behavior pp 123–54). Praeger, New York

Phillips CJ, Tandler B, Pinkstaff CA (1987) Unique salivary glands in two genera of tropical microchiropteran bats: an example of evolutionary convergence in histology and histochemistry. J Mammal 68(2):235–242

Popper AN, Fay RR (1995) Hearing by bats, vol 5. Springer, New York

Popper AN, Fay RR (2005) Sound source localization. Springer, New York

Poussin C, Simmons JA (1982) Low-frequency hearing sensitivity in the echolocating bat, Eptesicus fuscus. J Acoust Soc Am 72(2):340–342

Prado CPA, Haddad CF (2003) Testes size in leptodactylid frogs and occurrence of multimale spawning in the genus Leptodactylus in Brazil. J Herpetol 37(2):354–362

Puechmaille SJ, Borissov IM, Zsebok S, Allegrini B, Hizem M, Kuenzel S, Schuchmann M, Teeling EC, Siemers BM (2014) Female mate choice can drive the evolution of high frequency echolocation in bats: a case study with Rhinolophus mehelyi. PLoS ONE 9(7):e103452

Ratcliffe JM, Raghuram H, Marimuthu G, Fullard JH, Fenton MB (2005) Hunting in unfamiliar space: echolocation in the Indian false vampire bat, Megaderma lyra, when gleaning prey. Behav Ecol Sociobiol 58:157–164

Rhebergen F, Taylor RC, Ryan MJ, Page RA, Halfwerk W (2015) Multimodal cues improve prey localization under complex environmental conditions. Proc Royal Soc B: Biol Sci 282(1814)

Römer H, Lang A, Hartbauer M (2010) The signaller's dilemma: a cost–benefit analysis of public and private communication. PLoS ONE 5(10):e13325

Ron SR (2008) The evolution of female mate choice for complex calls in tungara frogs. Anim Behav 76(6):1783–1794

Rübsamen R, Neuweiler G, Sripathi K (1988) Comparative collicular tonotopy in two bat species adapted to movement detection, Hipposideros speoris and Megaderma lyra. J Comp Physiol A: Neuroethol Sens Neural Behav Physiol 163(2):271–285

Ruczyński I, Kalko EKV, Siemers BM (2009) Calls in the forest: a comparative approach to how bats find tree cavities. Ethology 115(2):167–177

Ryan MJ (1983) Sexual selection and communication in a Neotropical frog, Physalaemus pustulosus. Evolution 37(2):261–272

Ryan MJ (1986) Factors influencing the evolution of acoustic communication—biological constraints. Brain Behav Evol 28(1–3):70–82

Ryan MJ (1990) Sexual selection, sensory systems and sensory exploitation. Oxf Surv Evol Biol 7:157–195

Ryan MJ (2005) The evolution of behaviour, and integrating it towards a complete and correct understanding of behavioural biology. Anim Biol 55(4):419–439

Ryan MJ (2011) Sexual selection: a tutorial from the tungara frog. In: Losos JB (ed) In the light of evolution: essays from the laboratory and field. Roberts and Company, Greenwood Village CO, pp 185–203

Ryan MJ, Rand AS (2003) Mate recognition in tungara frogs: a review of some studies of brain, behavior, and evolution. Acta Zoologica Sinica 49(6):713–726

Ryan MJ, Tuttle MD (1983) The ability of the frog-eating bat to discriminate among novel and potentially poisonous frog species using acoustic cues. Anim Behav 31:827–833

Ryan MJ, Tuttle MD (1987) The role of prey-generated sounds, vision, and echolocation in prey localization by the African bat Cardioderma cor (Megadermatidae). J Comp Physiol A: Neuroethol Sens Neural Behav Physiol 161(1):59–66

Ryan MJ, Tuttle MD, Taft LK (1981) The costs and benefits of frog chorusing behavior. Behav Ecol Sociobiol 8(4):273–278

Ryan MJ, Tuttle MD, Rand AS (1982) Bat predation and sexual advertisement in a Neotropical anuran. Am Nat 119(1):136–139

Ryan MJ, Tuttle MD, Barclay RMR (1983) Behavioral responses of the frog-eating bat, Trachops cirrhosus, to sonic frequencies. J Comp Physiol 150(4):413–418

Sales G (2012) Ultrasonic communication by animals. Chapman & Hall, London

Schmidt S, Hanke S, Pillat J (2000) The role of echolocation in the hunting of terrestrial prey—new evidence for an underestimated strategy in the gleaning bat, Megaderma lyra. J Comp Physiol A: Neuroethol Sens Neural Behav Physiol 186(10):975–988

Schmidt S, Yapa W, Grunwald J-E (2010) Echolocation behaviour of *Megaderma lyra* during typical orientation situations and while hunting aerial prey: a field study. J Comp Physiol A: Neuroethol Sens Neural Behav Physiol 97(5):403–412

Schnitzler H-U, Kalko EKV (2001) Echolocation by insect-eating bats. Bioscience 51(7):557–569

Schnitzler H-U, Moss CF, Denzinger A (2003) From spatial orientation to food acquisition in echolocating bats. Trends Ecol Evol 18(8):386–394

Schuchmann M, Siemers BM (2010) Behavioral evidence for community wide species discrimination from echolocation calls in bats. Am Nat 176:72–82

Schuchmann M, Puechmaille SJ, Siemers BM (2012) Horseshoe bats recognise the sex of conspecifics from their echolocation calls. Acta Chiropterol 14(1):161–166

Schwartz JJ, Bee MA, Tanner SD (2000) A behavioral and neurobiological study of the responses of gray treefrogs, *Hyla versicolor* to the calls of a predator, *Rana catesbeiana*. Herpetologica 27–37

Seamark EC, Bogdanowicz W (2002) Feeding ecology of the common slit-faced bat (*Nycteris thebaica*) in KwaZulu-Natal. S Afr Acta Chiropterol 4(1):49–54

Seeba F, Schwartz JJ, Bee MA (2010) Testing an auditory illusion in frogs: perceptual restoration or sensory bias? Anim Behav 79(6):1317–1328

Shen JX, Xu ZM, Feng AS, Narins PM (2011) Large odorous frogs (*Odorrana graminea*) produce ultrasonic calls. J Comp Physiol A: Neuroethol Sens Neural Behav Physiol 197(10):1027–1030

Shetty S, Sreepada KS (2013) Prey and nutritional analysis of *Megaderma lyra* guano from the west coast of Karnataka, India. Adv Biores 4(3):1–7

Siemers BM, Stilz P, Schnitzler H-U (2001) The acoustic advantage of hunting at low heights above water: behavioural experiments on the European 'trawling' bats *Myotis capaccinii*, *M. dasycneme* and *M. daubentonii*. J Exp Biol 204:3843–3854

Siemers BM, Kriner E, Kaipf I, Simon M, Greif S (2012) Bats eavesdrop on the sound of copulating flies. Curr Biol 22(14):R563–R564

Simmons JA, Stein RA (1980) Acoustic imaging in bat sonar: echolocation signals and the evolution of echolocation. J Comp Physiol A: Neuroethol Sens Neural Behav Physiol 135(1):61–84

Speakman J (1993) The evolution of echolocation for predation. Paper presented at the Symposia of the Zoological Society of London

Stoddard PK (1999) Predation enhances complexity in the evolution of electric fish signals. Nature 400(6741):254–256

Surlykke A, Jakobsen L, Kalko EK, Page RA (2013) Echolocation intensity and directionality of perching and flying fringe-lipped bats, *Trachops cirrhosus* (Phyllostomidae). Front Physiol 4

Suthers RA (1965) Acoustic orientation by fish-catching bats. J Exp Zool 158(3):319–347

Tandler B, Phillips CJ, Nagato T (1996) Histological convergent evolution of the accessory submandibular glands in four species of frog-eating bats. Eur J Morphol 34(3):163–168

Tandler B, Nagato T, Phillips CJ (1997) Ultrastructure of the unusual accessory submandibular gland in the fringe-lipped bat, *Trachops cirrhosus*. Anatomical Record 248(2):164–175

Thomas AJ, Jacobs DS (2013) Factors influencing the emergence times of sympatric insectivorous bat species. Acta Chiropterol 15(1):121–132

Trillo PA, Athanas KA, Goldhill DH, Hoke KL, Funk WC (2013) The influence of geographic heterogeneity in predation pressure on sexual signal divergence in an Amazonian frog species complex. J Evol Biol 26(1):216–222

Tuttle MD, Ryan MJ (1981) Bat predation and the evolution of frog vocalizations in the Neotropics. Science 214(4521):677–678

Tuttle MD, Ryan MJ (1982) The role of synchronized calling, ambient light, and ambient noise, in anti-bat-predator behavior of a treefrog. Behav Ecol Sociobiol 11(2):125–131

Tuttle MD, Taft LK, Ryan MJ (1981) Acoustical location of calling frogs by Philander opossums. Biotropica 13(3):233–234

Tuttle MD, Taft LK, Ryan MJ (1982) Evasive behavior of a frog in response to bat predation. Anim Behav 30(MAY):393–397

Usman KA, Habersetzer J, Subbaraj R, Gopalkrishnaswamy G, Paramanandam K (1980) Behaviour of bats during a lunar eclipse. Behav Ecol Sociobiol 7(1):79–81

Vaughan TA (1976) Nocturnal behavior of the African false vampire bat (*Cardioderma cor*). J Mammal 57(2):227–248

Wells KD (2007) The ecology and behavior of amphibians (1st ed). The University of Chicago Press, Chicago

Wells KD, Schwartz JJ (1982) The effect of vegetation on the propagation of calls in the Neotropical frog *Centrolenella fleischmanni*. Herpetologica 38:449–455

Wiley RH, Richards DG (1978) Physical constraints on acoustic communication in the atmosphere: implications for the evolution of animal vocalizations. Behav Ecol Sociobiol 3 (1):69–94

Zuk M, Kolluru GR (1998) Exploitation of sexual signals by predators and parasitoids. Q Rev Biol 73(4):415–438

Chapter 7
Synthesis and Future Research

Abstract Insects have obviously responded to bat predation by evolving a range of defences that are specific to bat predation. However, apart from rare examples of stealth echolocation, only one of which appears to meet the criteria of co-evolution; there is no indication that bats have responded reciprocally and specifically to the defences of their prey. However, with the advent of new technologies, more examples of co-evolved stealth echolocation may be uncovered. This requires an increase in both the geographic and taxonomic coverage of bat–insect interactions. This would include the auditory thresholds of insects and the diets of bats at a level that would allow the determination of whether bats are eating tympanate or tympanate prey. Systems that are likely to yield examples of co-evolution would include those that involve some kind of trade-off, for example, low intensity and/or low-frequency echolocation calls in bats. Bat species in the family Molossidae would be good candidates for this. It is imperative that investigations of co-evolution between bats and their prey are done within a phylogenetic framework, so that the ancestral character states of both groups can be identified and the timing of emergence of suspected co-evolving traits can be determined. For this, we need dated phylogenies with excellent taxonomic coverage for suspected co-evolving lineages.

7.1 Synthesis

Co-evolution is characterized by the tandem evolution of traits of at least two lineages such that the evolutionary changes in the trait of one lineages are in direct response to the evolutionary changes in traits of the other lineage and vice versa. Thus, identification of co-evolution requires that the responses of each lineage be both reciprocal and specific in dyads of lineages or, in the case of diffuse co-evolution, between multiple lineages (Chap. 1).

There is, as yet, no evidence of pairwise co-evolution, strictly defined (e.g. Futuyma and Slatkin 1983) in the interaction between bats and their prey. If a case can be made for co-evolution at all in bat–prey interactions, it would be as an

© The Author(s) 2016
D.S. Jacobs and A. Bastian, *Predator–Prey Interactions: Co-evolution between Bats and their Prey*, SpringerBriefs in Animal Sciences,
DOI 10.1007/978-3-319-32492-0_7

instance of diffuse co-evolution. However, even in this instance definitive support is lacking. A strong case can be made in support of insect hearing (especially that of moths), or at least some characteristics of it, as well as moth ultrasonic clicks having evolved in direct response to bat echolocation. The timing of the evolution of moth hearing, the restriction of moth hearing to families which arose after echolocation (Yack and Fullard 2000), the matched auditory sensitivity of insects in general to the echolocation frequencies of most bats, the regression of audition in insects released from bat predation, and the functional significance of moths clicks in relation to bat predation all point to a system evolved in direct response to bat predation. This satisfies the specificity criterion of co-evolution. Although there are many characteristics of the echolocation system (e.g. passive acoustic prey detection) and behaviour of bats that make bats less conspicuous to moths, it is far from clear that these have evolved in direct response to prey defences, thus evidence for reciprocity is equivocal. As pointed out in previous chapters, for almost all of these bat traits there are alternative explanations for their evolutionary origin, and it is nearly impossible to tease apart co-evolution from these other explanations, e.g. adaptation to different habitats or responses that overcome the limitations imposed by the physical properties of sound.

The only convincing evidence of reciprocal and specific bat responses to prey defences are therefore in systems where traits making bat echolocation less conspicuous to prey are also associated with trade-offs. Such trade-offs suggest that changes in bat echolocation, putatively in response to prey defences, are not adaptive in the sense that they make echolocation more effective but that they facilitate circumvention of prey defences. Currently, there is only one such instance, namely the use of low-intensity calls by the aerial hawking bat, *Barbastella barbastellus*, that preys predominantly on eared moths (Goerlitz et al. 2010a). The use of low-intensity calls in this species could not have evolved for more effective echolocation. Low-intensity calls are usually used by substrate gleaning bats or bats which hunt close to vegetation because they minimize the masking of target echoes by background clutter (Fig. 2.3). The use of such calls by an aerial hawker, which hunts in open space without clutter, results in a reduction of detection range, a serious limitation for an aerial hawker because they have to scan larger volumes of open space for prey. The advantage is that this bat detects moths before they are aware of the bat, allowing the bat to be a more effective predator on eared moths.

It is possible that the *Euderma maculatum* may have taken this a step further by combining low-intensity calls with frequencies below that at which moths are most sensitive. *Euderma maculatum* is an aerial hawker (Leonard and Fenton 1983; Wai-Ping and Fenton 1989; and also see Watkins 1977), feeds predominantly on moths (Watkins 1977; Wai-Ping and Fenton 1989), and uses calls of relatively low intensity (80–90 dB SPL at 10 cm vs. 121–125 dB SPL for other vespertilionids; Woodsworth et al. 1981; Jensen and Miller 1999; Holderied et al. 2005) and frequency (dominant frequency 9–12 kHz) which allows it to detect the moths echo before the moths become aware of the bat (Fullard and Dawson 1997). However, it is not known whether the moths it feeds on are eared, and it is equivocal whether the calls of this species is indeed an adaptive response to moths hearing. Nevertheless, the echolocation calls of this bat do not represent adaptations that would maximize resolution and detection distance. Here too, the low amplitude of its calls would result in decreased detection distance but in addition the low frequency would mean that the bat is restricted to relatively large prey, a double trade-off. Such trade-offs suggest that the anti-bat defences of its prey drives the evolution of these otherwise disadvantageous signal characteristics.

Another group of bats that may suffer from a similar trade-off but may be able to circumvent moth hearing because they use low-frequency calls are bats in the family Molossidae. Molossids that use very low frequencies include *Tadarida teniotis* (11 kHz, Zbinden and Zingg 1986), *T. australis* (12.6 kHz, Fullard et al. 1991) and *Otomops martiensseni* (9–12 kHz, Fenton et al. 2004). Their frequencies are much lower than those used by other open air foragers (e.g. *T. aegyptiaca*, 22 kHz, Schoeman and Jacobs 2008), restricting them to mostly large insects (Rydell and Yalden 1997) because of the rather weak echoes from small insects (Waters 2003). This is a serious trade-off since large insects comprise a small portion of the insect fauna (Janzen and Schoener 1968). Furthermore, these bats feed predominantly on moths (Jones and Rydell 1994; Rydell and Yalden 1997; Lumsden and Bennett 2005) and therefore this case suggests that the low frequencies, which do not improve the operation of echolocation, were selected because they were less audible to moths.

7.2 Future Research

7.2.1 Hearing and Avoiding Bats by Their Prey

Researching the audibility to moths of echolocation calls of the bat species mentioned above (e.g. species in the family Molossidae) and determining whether or not these bats take eared moths could yield some interesting discoveries. Similar investigations should be done on bat species using two-tone echolocation calls (Mora et al. 2014) but in addition, playback experiments should be conducted in which the intensity of the second call in the two-tone pairs of calls is increased to determine whether moths then respond to the calls of these bats. The first and second calls in such pairs have different dominant frequencies and moths may interpret such calls as having much longer intervals, leading to a misinterpretation of the distance to the bat and thus failing to execute its evasive flight behaviour to escape the bat (Chap. 5). Some molossids using low frequencies are known to also include small moths in their diets (Rydell and Yalden 1997; Lumsden and Bennett 2005), suggesting that some degree of behavioural foraging flexibility (e.g. slower flight speed and shorter duration calls) may allow these species to detect and capture insects smaller than that suggested by their low-frequency echolocation alone (Rydell and Yalden 1997). Here, field observations and prey capture performance experiments may elucidate multimodal capture methods used by bats in combination with echolocation.

The use of multiple cues has been demonstrated in the frog-eating bat *Trachops cirrhosus*. This species mainly uses the frogs' calls to detect the frogs but also uses its echolocation to detect the movements of the vocal sac and the ripples in the water made by movement of the vocal sac when the frog is calling. Such ripples remain detectable for a few seconds after the frog has ceased calling (Chap. 6). There has been very little research on frogs as prey of bats (with exception of the frog species eaten by *T. cirrhosus*) and there may be many novel adaptive defences of frogs awaiting discovery. Of special interest in this regard is the ability of some recently described frog species to emit and hear ultrasound (Chap. 6). Investigating if more frog species are capable of ultrasound perception (especially those species that are known to be preyed upon by bats) as well as studies on the responses of frogs to bat echolocation would improve our understanding of the predator–prey interactions between bats and frogs in the context of co-evolution.

7.2.2 Co-evolution Confounded by Sexual Selection

Although some authors (e.g. Krebs and Davies 1991; Holland and Rice 1999; Arnqvist and Rowe 2005) refer to sexual selection as an example of co-evolution between male traits and female preference for those traits, the two processes have glaring differences but can have similar consequences (Chap. 1). This could result in certain phenotypic patterns being erroneously ascribed to one process when the other or both are responsible. For example, the perception of the echoes from the spinning tails of *Luna moths* by bats would be a fruitful area of investigation especially if done in an evolutionary context including the role that sexual selection plays on the length of the tails. Sexual dimorphism in tail length is evident in *L. moths* (Barber et al. 2015) but since both genders have elongated tails, sexual selection cannot be solely responsible and may only be a contributory factor. This emphasizes that systems should be well understood and several hypotheses should be tested simultaneously before drawing conclusions about the processes responsible for the phenomena we observe.

A potential fruitful area of research would be the relative roles of sexual selection, natural selection, and co-evolution between predator and prey in bat systems in which there is pronounced sexual dimorphism, particularly in echolocation calls, accompanied by gender differences in diet. Sexual selection might give rise to sexual dimorphism if mate choice by females shows a preference for certain echolocation frequencies—a hypothesis discussed and tested by recent research (Knörnschild et al. 2012; Puechmaille et al. 2014). Natural selection may favour higher-nutritional diet for females while they lactate or while they are pregnant. Higher-nutritional diet may require different diets which require different echolocation frequencies. Teasing apart these two kinds of processes from co-evolution is essential if we are to understand their relative roles.

7.2.3 Gleaning Bats: Adaptation to Habitat
or Co-evolution?

The facility with which bats are able to detect and capture stationary prey from the substrate in dense vegetation has a direct influence on the effectiveness of prey defences. The extent to which bats can combine three-dimensional flight and echolocation (Geipel et al. 2013) to generate prey-specific acoustic images (Simmons and Stein 1980; Neuweiler 1990) allowing them to distinguish prey from a twig, for example, awaits ensonification experiments with prey in different background situations (Geipel et al. 2013).

However, some gleaners may be able to use the multiharmonic structure of their calls to suppress backward masking (i.e. overlap between echoes from the target and background) as has been reported in an aerial hawking bat that forages close to vegetation. The aerial-hawking big brown bat, *Eptesicus fuscus*, uses its second harmonic to separate echoes from the background from echoes from the target when foraging close to vegetation (Bates et al. 2011). Echoes reflected off objects located off the axis of the central beam of the bat's echolocation or further away than the target are weakened to a greater extent at the higher frequencies of the second harmonic than at the lower frequencies of the first harmonic. This is due to the increased distance of off-axis objects and the greater atmospheric attenuation of higher frequencies in the second harmonic. In contrast, echoes from the target are derived from interference between multiple, overlapping reflections that reinforce and cancel each other at specific frequencies. Thus, echoes of the second harmonic reflected off a nearby target will have similar amplitude to that of the first harmonic, whereas echoes of the second harmonic arriving from the sides or farther away will be weaker because of its higher frequency. The bats are able to exploit the relative weakness of the second harmonic in echoes to categorize clutter as irrelevant and minimize backward masking (Bates et al. 2011). Whether or not gleaning bats use the same strategy to minimize background masking (Fig. 2.3) remains to be discovered. If so, this strategy may represent a co-evolutionary response by bats to those insects which hide in dense vegetation to escape bat predation.

It has been suggested that the different results obtained for different gleaners with respect to their reliance on echolocation to locate prey in clutter are an artefact of the experimental set-up e.g. associative learning during the experiment and the equipment used (Ratcliffe et al. 2005). If so, experiments on naïve bats hunting naturally in the wild (e.g. Schmidt et al. 2000; Corcoran and Conner 2012) are sorely needed to determine whether, to what extent, and how different gleaning bat species use echolocation to hunt in clutter.

The use of harmonics, as described for *E. fuscus*, by gleaning bats and the combination of echolocation and three-dimensional flight would represent traits that have evolved in direct response to insects that use backward masking effects to hide from bats. If so, it may provide evidence of a response in the echolocation of bats that is independent of habitat effects.

7.2.4   Flight Patterns and Echolocation Behaviour
as Response to Prey Defence Strategies
and the Importance of Knowing the Bats' Diet

For many bat species, responses to insect hearing may be behavioural. Behavioural flexibility in flight and echolocation may allow bats whose calls are audible to moths to counter, if not the hearing, at least the acoustic startle (ASR) or freezing response of some insects. This could include slower flight for more reaction time when dealing with insects that generate only weak echoes (Rydell and Yalden 1997) or approaching from a particular angle or some other flight manoeuvre to counter ASRs (Conner and Corcoran 2012). For example, *Lasiurus borealis* attacks prey from below thus circumventing moths that dive when they become aware of bats (Conner and Corcoran 2012). Diving is normally a more successful moth strategy than flying away (Corcoran and Conner 2012).

Echolocation behaviour of bats is as flexible, or even more so, than flight and the details of how bats vary the temporal and spectral components and intensities (=detectability) of their echolocation when pursuing eared prey versus deaf prey is likely to provide us with insight into the versatility of echolocation in dealing with insect defences.

More importantly, the details of the diets of most bats are unknown. These details include species or at least family of prey taken, size of prey, seasonal variation in diet, whether prey are eared or not, where bats hunt particular prey, and the detectability of insect prey. Thoughtful use of barcoding methods will make it easier to analyse the diets of bats. The setting of bat–prey interactions may have much influence on the outcome of such interactions. It also remains to be tested whether prey are detectable and ignored because they are unprofitable or unavailable because they are undetectable. Studies done so far have not been able to test between these two explanations (Houston et al. 2004) but it is likely that both of these explanations are correct. Whether the low proportions of insects that are camouflaged by shape (at least to most bats) in the diets of these bats is due to their low proportion in the prey fauna or because they are nevertheless still more cryptic than other prey, even to bats such as *Micronycteris microtis*, remains to be tested. This is an exciting area of research that may yield novel discoveries such as a predator–prey arms race involving acoustic crypsis mediated by selection on echolocation and the shape, texture and material of prey bodies.

7.2.5   Some Interesting Areas of Research
on Earless Bat Prey

Part of the problem of identifying whether bat adaptations are in response to insect defences and vice versa is that the timing of the origin of the traits of interest in both groups involved in the interaction are mostly unknown. It is imperative that investigations of co-evolution between bats and their prey are done within a phylogenetic framework so that the ancestral character states of both groups can be identified. Such knowledge will advance our understanding of the evolution of the interaction been bats and their prey much more rapidly than without. More generally, one would need to aim for a comprehensive approach which investigates Tinbergens' (1963) four facets of the 'why' question: causation (a proximate explanation), development/ontogeny (proximate), function/adaptive value (ultimate), and phylogeny/history (ultimate).

Many earless insects lose reproductive opportunities as result of becoming inactive, e.g. reducing flying time, to avoid bats. However, reduced activity poses problems for organisms that are short lived and depend on flight for reproductive opportunities. Investigations into how earless insects circumvent these constraints imposed upon them by their bat-avoidance strategies are sorely needed. Questions that are worth addressing include the following: do earless moths have longer lifespans than their eared relatives, do males have antennae that are more sensitive (i.e. relatively larger) to pheromones for more efficient (and therefore quicker) localization of females (Soutar and Fullard 2004), do earless moths avoid bats by using short bouts of fast, and erratic flight when they are active (Roeder 1974)?

Given the widespread occurrence of flightlessness among bat prey, an investigation into the role that bats played in the evolution of flightlessness would benefit from a comparative ecological and phylogenetic approach. Questions that need to be answered are what are the ecological correlates (e.g. in terms of habitat or the sympatric bat community) of flightlessness, is flightlessness always accompanied by increased fecundity, and what is the distribution and timing of flightlessness across the phylogenies of bat prey?

In response to the simulated bat calls, males of the two species of moths, *Pseudaletia unipuncta* (Noctuidae) and *Ostrinia nubilalis* (Pyralidae), aborted their upwind flight in a pheromone plume and females stopped releasing pheromone (Acharya and McNeil 1998). It would be interesting to determine how the evolution of female cessation of pheromone release is co-ordinated with male cessation of flight. Such co-ordination is needed because any female that did not co-ordinate pheromone release with male activity would be wasting resources. Could the genes for male cessation of flight and female cessation of pheromone release be linked as in female choice sexual selection? This could provide linkage between the processes of sexual selection and co-evolution.

7.2.6 Improving the Taxonomic and Geographic Coverage of Bat–Prey Interactions

Our understanding of the bat–prey interaction is sorely limited by the very narrow taxonomic and geographic coverage of both prey and predator. For example, we know much about anti-bat defences in moths and crickets and comparatively almost nothing in other insects. Similarly, we know very little about the interactions between bats and their vertebrate prey and almost all that is known is on the frog species preyed upon by *T. cirrhosus*. Furthermore, almost all of our knowledge of bat-eating frogs comes from just two species; *T. cirrhosis* (predominantly) and *Megaderma lyra*. Comparatively little research on any prey or bat taxa has been done outside the Americas and Europe. This is attributable to the technologically advanced, expensive equipment and specialized skills needed to do the necessary research. However, we have to find ways of remedying this, particularly in areas outside of the Americas and Europe where there are few biologists working on sensory ecology but where the diverse fauna are likely to hide many novel discoveries. We have to improve our taxonomic and geographic coverage of the sensory ecology of both predator and prey if we hope to understand all the processes, including co-evolution that has generated the enormous diversity we observe.

7.2.7 Life-Dinner Principle

Finally, it is to be expected that co-evolution is likely to be a rare event in any predator–prey system, largely because both predator and prey are exposed to multiple, and at times opposing, selection pressures. The ability of each to respond to the other is therefore limited. However, the greatest limitation to the evolution of reciprocal and specific traits in predator and prey is that the selective pressures exerted on the players in the interaction are not equal. As the life-dinner principle suggests (Dawkins and Krebs 1979), selection pressure exerted on the prey is much greater in magnitude than that exerted on predator. One would therefore expect a much greater response from the prey than from the predator. This imbalance ultimately limits reciprocity, consigning co-evolution to a rare but nevertheless important evolutionary phenomenon that has amazed us for centuries.

References

Acharya L, McNeil JN (1998) Predation risk and mating behavior: the responses of moths to bat-like ultrasound. Behav Ecol 9(6):552–558

Arnqvist G, Rowe L (2005) Sexual conflict. Princeton University Press, New Jersey

Barber JR, Leavell BC, Keener AL, Breinholt JW, Chadwell BA, McClure CJW, Hill GM, Kawahara AY (2015) Moth tails divert bat attack: evolution of acoustic deflection. Proc Natl Acad Sci USA 112(9):2812–2816

Bates ME, Simmons JA, Zorikov TV (2011) Bats use echo harmonic structure to distinguish their targets from background clutter. Science 333(6042):627–630

Conner WE, Corcoran AJ (2012) Sound strategies: The 65-million-year-old battle between bats and insects. In: Berenbaum MR (ed) Annual review of entomology, vol 57. Annual Reviews, Palo Alto, pp 21–39

Corcoran AJ, Conner WE (2012) Sonar jamming in the field: effectiveness and behavior of a unique prey defense. J Exp Biol 215(24):4278–4287

Dawkins R, Krebs JR (1979) Arms races between and within species. Proc Royal Soc Lond B: Biol Sci 205(1161):489–511

Fenton MB, Jacobs DS, Richardson EJ, Taylor PJ, White E (2004) Individual signatures in the frequency-modulated sweep calls of African large-eared, free-tailed bats Otomops martiensseni (Chiroptera: Molossidae). J Zool 262:11–19

Fullard JH, Dawson JW (1997) The echolocation calls of the spotted bat Euderma maculatum are relatively inaudible to moths. J Exp Biol 200(1):129–137

Fullard JH, Koehler C, Surlykke A, McKenzie NL (1991) Echolocation ecology and flight morphology of insectivorous bats (Chiroptera) in South-western Australia. Aust J Zool 39 (4):427–438

Futuyma DJ, Slatkin M (1983) Introduction. In: Coevolution. Sinauer Associates Inc, Sunderland

Geipel I, Jung K, Kalko EKV (2013) Perception of silent and motionless prey on vegetation by echolocation in the gleaning bat Micronycteris microtis. Proc Royal Soc Lond B: Biol Sci 280 (1754):7

Goerlitz HR, ter Hofstede HM, Zeale MR, Jones G, Holderied MW (2010) An aerial-hawking bat uses stealth echolocation to counter moth hearing. Curr Biol 20(17):1568–1572

Holderied MW, Korine C, Fenton MB, Parsons S, Robson S, Jones G (2005) Echolocation call intensity in the aerial hawking bat Eptesicus bottae (Vespertilionidae) studied using stereo videogrammetry. J Exp Biol 208(7):1321–1327

Holland B, Rice WR (1999) Experimental removal of sexual selection reverses intersexual antagonistic coevolution and removes a reproductive load. Proc Natl Acad Sci USA 96 (9):5083–5088

Houston RD, Boonman AM, Jones G (2004) Do echolocation signal parameters restrict bats' choice of prey? Echolocation in bats and dolphins. University of Chicago Press, Chicago, pp 339–345

Janzen DH, Schoener TW (1968) Differences in insect abundance and diversity between wetter and drier sites during a tropical dry season. Ecology 96–110

Jensen ME, Miller LA (1999) Echolocation signals of the bat Eptesicus serotinus recorded using a vertical microphone array: effect of flight altitude on searching signals. Behav Ecol Sociobiol 47:60–69

Jones G, Rydell J (1994) Foraging strategy and predation risk as factors influencing emergence time in echolocating bats. Philos Transac Royal Soc B: Biol Sci 346(1318):445–455

Knörnschild M, Jung K, Nagy M, Metz M, Kalko E (2012) Bat echolocation calls facilitate social communication. Proc Royal Soc London B: Biol Sci 279(1748):4827–4835

Krebs JR, Davies NB (1991) Behavioural ecology: an evolutionary approach, 3rd edn. Blackwell Science Ltd., Oxford

Leonard ML, Fenton MB (1983) Habitat use by spotted bats (Euderma maculatum, Chiroptera, Vespertilionidae)—roosting and foraging behavior. Can J Zool 61(7):1487–1491

Lumsden LF, Bennett AF (2005) Scattered trees in rural landscapes: foraging habitat for insectivorous bats in south-eastern Australia. Biol Conserv 122(2):205–222

Mora EC, Fernández Y, Hechavarría J, Pérez M (2014) Tone-deaf ears in moths may limit the acoustic detection of two-tone bats. Brain Behav Evol 83(4):275–285

Neuweiler G (1990) Auditory adaptations for prey capture in echolocating bats. Physiol Rev 70 (3):615–641

Puechmaille SJ, Borissov IM, Zsebok S, Allegrini B, Hizem M, Kuenzel S, Schuchmann M, Teeling EC, Siemers BM (2014) Female mate choice can drive the evolution of high frequency echolocation in bats: a case study with Rhinolophus mehelyi. PLoS ONE 9(7):e103452

Ratcliffe JM, Raghuram H, Marimuthu G, Fullard JH, Fenton MB (2005) Hunting in unfamiliar space: echolocation in the Indian false vampire bat, Megaderma lyra, when gleaning prey. Behav Ecol Sociobiol 58:157–164

Roeder KD (1974) Acoustic sensory responses and possible bat-evasion tactics of certain moths. Paper presented at the proceedings of the Canadian Society of zoologists annual meeting

Rydell J, Yalden DW (1997) The diets of two high-flying bats from Africa. J Zool 242:69–76
Schmidt S, Hanke S, Pillat J (2000) The role of echolocation in the hunting of terrestrial prey—
    new evidence for an underestimated strategy in the gleaning bat, *Megaderma lyra*. J Comp
    Physiol A: Neuroethol Sens Neural Behav Physiol 186(10):975–988
Schoeman MC, Jacobs DS (2008) The relative influence of competition and prey defenses on the
    phenotypic structure of insectivorous bat ensembles in southern Africa. PLoS ONE 3(11):
    e3715
Simmons JA, Stein RA (1980) Acoustic imaging in bat sonar—echolocation signals and the
    evolution of echolocation. J Comp Physiol 135(1):61–84
Soutar AR, Fullard JH (2004) Nocturnal anti-predator adaptations in eared and earless Nearctic
    Lepidoptera. Behav Ecol 15(6):1016–1022
Tinbergen N (1963) On aims and methods of ethology. Zeitschrift für Tierpsychologie 20
    (4):410–433
Wai-Ping V, Fenton MB (1989) Ecology of spotted bat (*Euderma maculatum*) roosting and
    foraging behavior. J Mammal 70(3):617–622
Waters DA (2003) Bats and moths: what is there left to learn? Physiol Entomol 28(4):237–250
Watkins LC (1977) *Euderma maculatum*. Mamm Species 77:1–4
Woodsworth G, Bell G, Fenton M (1981) Observations of the echolocation, feeding behaviour,
    and habitat use of *Euderma maculatum* (Chiroptera: Vespertilionidae) in southcentral British
    Columbia. Can J Zool 59(6):1099–1102
Yack JE, Fullard JH (2000) Ultrasonic hearing in nocturnal butterflies. Nature 403(6767):265–266
Zbinden K, Zingg PE (1986) Search and hunting signals of echolocating European free-tailed bats,
    Tadarida teniotis, in southern Switzerland. Mammalia 50(1):7–25

Glossary

Acoustic aposematism The use of an acoustic signal by the prey to warn predators of their unpalatability

Acoustic fovea An area of high sensitivity to a specific, narrow range of frequencies in the auditory pathway

Acoustic glints Changes in the amplitude and frequency in the echoes of bat calls from flapping insect wings perceived by the bat as amplitude and frequency glints against the constant echo from the background clutter

Acoustic startle response A change in behaviour, usually involving a cessation of activity, in response to acoustic stimuli produced by a predator

Aerial hawking Pursuit and capture of aerial prey while in flight

Allotonic Sound frequencies outside of the range of frequencies under consideration

Backward masking Overlap of background echoes with echoes from the target making it difficult for echolocating bats that do not use acoustic glints to detect the echoes from the target

Batesian mimicry A predator avoidance strategy in which palatable prey mimic unpalatable prey to benefit from the warning signal of the unpalatable prey

Cerci Paired appendages on the rear-most segments of many arthropods

Chordotonal organ A stretch receptor organ in insects and other arthropods responsive to distention of various organs and muscles allowing the animal to keep track of the position of different body parts

Chorus A group of animals calling/singing together

Clutter In the context of echolocation, it is the echoes from non-target objects e.g. background vegetation

© The Author(s) 2016
D.S. Jacobs and A. Bastian, *Predator–Prey Interactions: Co-evolution between Bats and their Prey*, SpringerBriefs in Animal Sciences,
DOI 10.1007/978-3-319-32492-0

Co-evolution An iterative process among two or more lineages in which traits in each lineage have evolved directly in response to traits in the other lineage (specificity) and each lineage has also responded to the evolution of the other (reciprocity)

Crypsis The minimization of detection through the use of visual, chemical, tactile, electric and acoustic traits when potentially detectable by an observer

Diel A period of 24 h (=daily)

Dilution effect The decrease in the probability of an individual being taken by a predator as a consequence of group size

Doppler shift compensation The emission of echolocation calls at frequencies lower than the hearing sensitivity of the acoustic fovea so that the upward shift in frequency of the echo, as a result of the relative velocity of the bat to the target, ensures that frequency of the echo is within frequency range of the acoustic fovea

Duty cycle The ratio of the duration of the call to the period of the call (call duration + time to the next call). Usually expressed as a percentage

Forward masking Overlap of the emitted echolocation call with the echo reflected from a target because the target is too close to the bat making. The echo is said to be masked for low duty cycle bats because the bats hearing is still "switched off" to avoid self-deafening

Gleaning Catching prey from the substrate

Lek An aggregation of males gathered to perform competitive displays to attract and compete for females

Muellerian mimicry A predator avoidance strategy in which unpalatable prey mimic other unpalatable prey as a collective warning signal to the predator of their unpalatability

Primary defence Defensive strategies used by prey that prevent detection by predators

Proprioceptors Sensory receptor which receives internal stimuli, especially one that responds to position and movement

Secondary defence Defensive strategies used by prey that improve survival after detection by predators

Sensory bias A pre-existing bias in the senses of one sex towards certain stimuli, such bias having evolved in a non-mating context. This bias is then exploited by the other sex in a mating context to obtain more mating opportunities

Sexual selection A mode of natural selection operating on traits directly involved in reproduction in the context of mate choice and competition for mates

Stridulate The generation of sound in insects by rubbing legs, wings or other body parts together

Tremulation Vibrations generated by unspecialized parts of the body (e.g. oscillations of the abdomen) and usually transmitted via legs through the substrate on which the insect is standing

Tympanate Possessing a tympanic membrane

Tympanic membrane A thin membrane that separates the external ear from the middle ear and converts air pressure fluctuations caused by the transmission of sound to vibrations

Vocal sac The flexible membrane of skin in male frogs used to amplify their mating/advertisement calls

Bibliography

Anderson MG, Ross HA, Brunton DH, Hauber ME (2009) Begging call matching between a specialist brood parasite and its host: a comparative approach to detect co-evolution. Biol J Linnean Soc 98(1):208–216

Aviles JM, Vikan JR, Fossoy F, Antonov A, Moksnes A, Roskaft E, Stokke BG (2010) Avian colour perception predicts behavioural responses to experimental brood parasitism in chaffinches. J Evol Biol 23(2):293–301

Bán M, Moskát C, Barta Z, Hauber ME (2013) Simultaneous viewing of own and parasitic eggs is not required for egg rejection by a cuckoo host. Behav Ecol 24(4):1014–1021

Bates HW (1862) Contributions to an Insect Fauna of the Amazon Valley. Lepidoptera: Heliconidae. Trans Linnean Soc Lond 23(3):495–566.

Davies NB, Brooke MD (1988) Cuckoos versus reed warblers—adaptations and counter adaptations. Animal Behav 36:262–284

Davies NB, Brooke MD (1989) An experimental-study of co-evolution between the cuckoo, *Cuculus canorus*, and its hosts host egg discrimination. J Animal Ecol 58(1):207–224

De Mársico MC, Gantchoff MG, Reboreda JC (2012) Host–parasite co-evolution beyond the nestling stage? Mimicry of host fledglings by the specialist screaming cowbird. Proc Royal Soc Lond B: Biol Sci 279(1742):3401–3408

Grim T (2007) Experimental evidence for chick discrimination without recognition in a brood parasite host. Proc Royal Soc Lond B: Biol Sci 274(1608):373–381

Herkt KMB, Barnikel G, Skidmore AK, Fahr J (2016) A high-resolution model of bat diversity and endemism for continental Africa. Ecol Modell 320:9–28

Honza M, Prochazka P, Stokke BG, Moksnes A, Roskaft E, Capek M, Mrlik V (2004) Are blackcaps current winners in the evolutionary struggle against the common cuckoo? J Ethol 22 (2):175–180

Hosoi SA, Rothstein SI (2000) Nest desertion and cowbird parasitism: evidence for evolved responses and evolutionary lag. Animal Behav 59:823–840

Humphrie DA, Driver PM (1970) Protean defence by prey animals. Oecologia 5(4):285–302.

Krueger O, Davies NB (2002) The evolution of cuckoo parasitism: a comparative analysis. Proc Royal Soc Lond B: Biol Sci 269(1489):375–381

Langmore NE, Hunt S, Kilner RM (2003) Escalation of a co-evolutionary arms race through host rejection of brood parasitic young. Nature 422(6928):157–160

Langmore NE, Kilner RM, Butchart SHM, Maurer G, Davies NB, Cockburn A, MacGregor NA, Peters A, Magrath MJL, Dowling DK (2005) The evolution of egg rejection by cuckoo hosts in Australia and Europe. Behav Ecol 16(4):686–692

Langmore NE, Maurer G, Adcock GJ, Kilner RM (2008) Socially acquired host-specific mimicry and the evolution of host races in Horsfield's bronze-cuckoo *Chalcites basalis*. Evolution 62 (7):1689–1699

Langmore NE, Spottiswoode CN (2012) Visual trickery in avian brood parasites. Oxford University Press, Oxford

© The Author(s) 2016
D.S. Jacobs and A. Bastian, *Predator–Prey Interactions: Co-evolution between Bats and their Prey*, SpringerBriefs in Animal Sciences,
DOI 10.1007/978-3-319-32492-0

Langmore NE, Stevens M, Maurer G, Kilner RM (2009) Are dark cuckoo eggs cryptic in host nests? Animal Behav 78(2):461–468

Lazure L, Fenton MB (2011) High duty cycle echolocation and prey detection by bats. J Exp Biol 214(7):1131–1137

Marchetti K (2000) Egg rejection in a passerine bird: size does matter. Animal Behav 59:877–883

Mason P, Rothstein SI (1986) Co-evolution and avian brood parasitism: cowbird eggs show evolutionary response to host discrimination. Evolution 40(6):1207–1214

Moksnes A, Roskaft E, Braa AT (1991) Rejection behavior by common cuckoo hosts towards artificial brood parasite eggs. Auk 108(2):348–354

Roff DA (1990) The evolution of flightlessness in insects. Ecol Monogr 60(4):389–421

Sato NJ, Tokue K, Noske RA, Mikami OK, Ueda K (2010) Evicting cuckoo nestlings from the nest: a new anti-parasitism behaviour. Biol Lett 6(1):67–69

Schmidt S (1992) Perception of structured phantom targets in the echolocating bat, *Megaderma lyra*. J Acoust Soc Am 91(4):2203–2223

Schmidt S, Hanke S, Pillat J (2000) The role of echolocation in the hunting of terrestrial prey-new evidence for an underestimated strategy in the gleaning bat, *Megaderma lyra*. J Comp Physiol A: Neuroethol Sens Neural Behav Physiol 186(10):975–988

Spottiswoode CN, Stevens M (2010) Visual modeling shows that avian host parents use multiple visual cues in rejecting parasitic eggs. Proc Nat Acad Sci U.S.A. 107(19):8672–8676

Spottiswoode CN, Stevens M (2011) How to evade a co-evolving brood parasite: egg discrimination versus egg variability as host defences. Proc Royal Soc Lond B: Biol Sci 278(1724):3566–3573

Stoddard MC, Kilner RM, Town C (2014) Pattern recognition algorithm reveals how birds evolve individual egg pattern signatures. Nat Commun 5:10

Stoddard MC, Stevens M (2010) Pattern mimicry of host eggs by the common cuckoo, as seen through a bird's eye. Proc Royal Soc Lond B: Biol Sci 277(1686):1387–1393

Stoddard MC, Stevens M (2011) Avian vision and the evolution of egg color mimicry in the common cuckoo. Evolution 65(7):2004–2013

Stokke BG, Moksnes A, Roskaft E (2002) Obligate brood parasites as selective agents for evolution of egg appearance in passerine birds. Evolution 56(1):199–205

Tokue K, Ueda K (2010) Mangrove Gerygones Gerygone laevigaster eject Little Bronze-cuckoo *Chalcites minutillus* hatchlings from parasitized nests. Ibis 152(4):835–839

Triblehorn JD, Schul J (2009) Sensory-encoding differences contribute to species-specific call recognition mechanisms. J Neurophysiol 102(3):1348–1357

Welbergen J, Komdeur J, Kats R, Berg M (2001) Egg discrimination in the Australian reed warbler (*Acrocephalus australis*): rejection response toward model and conspecific eggs depending on timing and mode of artificial parasitism. Behav Ecol 12(1):8–15

Index

© The Author(s) 2016   123
D.S. Jacobs and A. Bastian, *Predator–Prey Interactions: Co-evolution between
Bats and their Prey*, SpringerBriefs in Animal Sciences,
DOI 10.1007/978-3-319-32492-0

Index of Species

© The Author(s) 2016

D.S. Jacobs and A. Bastian, *Predator–Prey Interactions: Co-evolution between Bats and their Prey*, SpringerBriefs in Animal Sciences,
DOI 10.1007/978-3-319-32492-0

order in text	Index of taxa as they appear in the text	other taxonomic classes	Class	Order	Family	Genus	Species	Common name
11	Birds		Aves					
10	Fish		Fish					
108	Butterfly family Hedyloidea	Hedyloidea	Insect	Lepidoptera	Hedylidae			
103	Noctuoidea	Noctuoidea	Insect	Lepidoptera				
102	Papilionoidea	Papilionoidea	Insect	Lepidoptera				
101	Pyraloidea	Pyraloidea	Insect	Lepidoptera				
33	Isoptera		Insect	Blattodea				
42	Cockroaches		Insect	Blattodea				
28	Cicindelidae		Insect	Coleoptera	Carabidae			
27	Scarabaeidae		Insect	Coleoptera	Scarabaeidae			
26	Coleoptera		Insect	Coleoptera				
29	Diptera		Insect	Diptera				
31	Ephemeroptera		Insect	Ephemeroptera				
54	Water strider, Aquarius naja, Hemiptera: Gerridae		Insect	Hemiptera	Gerridae	Aquarius	Aquarius naja	Water strider, River skater
34	Heteroptera		Insect	Hemiptera				
73	Balboa tibialis, Pseudophyllinae		Insect	Hemiptera	Tettigoniidae (=katydids)	Balboa	Balboa tibialis	
111	Lasiocampidae		Insect	Lepidoptera	Lasiocampidae			
96	Tiger moth, Bertholdia trigona		Insect	Lepidoptera	Arctiidae	Bertholdia	Bertholdia trigona	Tiger moth
62	Dogbane tiger moth, Cycnia tenera		Insect	Lepidoptera	Arctiidae	Cycnia	Cycnia tenera	Dogbane tiger moth
95	Arctiin moth, Cycnia tenera		Insect	Lepidoptera	Arctiidae	Cycnia	Cycnia tenera	
97	Moth, Syntomeida epilais		Insect	Lepidoptera	Arctiidae	Syntomeida	Syntomeida epilais	
65	Moths, Ostrinia nubilalis		Insect	Lepidoptera	Crambidae	Ostrinia	Ostrinia nubilalis	
98	Moth, Euchaetes egle		Insect	Lepidoptera	Erebidae	Euchaetes	Euchaetes egle	
109	Gypsy moths, Lymantria dispar		Insect	Lepidoptera	Erebidae	Lymantria	Lymantria dispar	Gypsy moths

(continued)

(continued)

(continued)

(continued)

order in text	Index of taxa as they appear in the text	other taxonomic classes	Class	Order	Family	Genus	Species	Common name
14	Crickets		Insect	Orthoptera	Gryllidae			
70	Creosote bush katydid, Insara covilleae		Insect	Orthoptera	Tettigoniidae	*Insara*	*Insara covilleae*	Creosote bush katydid
71	Nearctic katydid, Neoconocephalus ensiger		Insect	Orthoptera	Tettigoniidae	*Neoconocephalus*	*Neoconocephalus ensiger*	Nearctic katydid
84	Ground cricket, Eunemobius carolinus, Nemobiinae		Insect	Orthoptera	Trigonidiidae	*Eunemobius*	*Eunemobius carolinus*	Ground cricket
87	Migratory locust, Locusta migratoria, Acrididae		Insect	Orthoptera	Acrididae	*Locusta*	*Locusta migratoria*	Migratory locust
93	Teleogryllus oceanicus		Insect	Orthoptera	Gryllidae	*Teleogryllus*	*Teleogryllus oceanicus*	
58	Pacific field cricket, Teleogryllus oceanicus, Orthoptera: Gryllidae		Insect	Orthoptera	Gryllidae	*Teleogryllus*	*Teleogryllus oceanicus*	Pacific field cricket
69	Mole crickets, Gryllotalpidae		Insect	Orthoptera	Gryllotalpidae			Mole crickets
82	phaneropterines katydid, Amblycorypha oblongifolia		Insect	Orthoptera	Tettigoniidae (=katydids)	*Amblycorypha*		
92	Katydid, Copiphora brevirostris		Insect	Orthoptera	Tettigoniidae (=katydids)	*Copiphora*	*Copiphora brevirostris*	
75	Pseudophylline, Docidocercus gigliotosi		Insect	Orthoptera	Tettigoniidae (=katydids)	*Docidocercus*	*Docidocercus gigliotosi*	
74	Katydid Ischnomela gracilis, Pseudophyllinae		Insect	Orthoptera	Tettigoniidae (=katydids)	*Ischnomela*	*Ischnomela gracilis*	
80	copiphorine katydid, N. affinis		Insect	Orthoptera	Tettigoniidae (=katydids)	*Neoconocephalus*	*Neoconocephalus affinis*	
85	Tettigoniid, Neoconocephalus ensiger		Insect	Orthoptera	Tettigoniidae (=katydids)	*Neoconocephalus*	*Neoconocephalus ensiger*	

(continued)

(continued)

order in text	Index of taxa as they appear in the text	other taxonomic classes	Class	Order	Family	Genus	Species	Common name
89	Long-winged katydid, Phaneroptera falcata		Insect	Orthoptera	Tettigoniidae (=katydids)	Phaneroptera	Phaneroptera falcata	Long-winged katydid
79	Katydid, Scopiorinus fragilis, Tettigoniidae		Insect	Orthoptera	Tettigoniidae (=katydids)	Scopiorinus	Scopiorinus fragilis	
81	phaneropterines katydid, Steirodon rufolineatum		Insect	Orthoptera	Tettigoniidae (=katydids)	Scopiorinus	Steirodon rufolineatum	
76	Tettigoniids, Tettigonia cantans		Insect	Orthoptera	Tettigoniidae (=katydids)	Tettigonia	Tettigonia cantans	
86	Tettigonia viridissima		Insect	Orthoptera	Tettigoniidae (=katydids)	Tettigonia	Tettigonia viridissima	
77	Tettigoniids, Tettigonia viridissima		Insect	Orthoptera	Tettigoniidae (=katydids)	Tettigonia		
15	Katydids		Insect	Orthoptera	Tettigoniidae (=katydids)		Neoconocephalus retusus	
40	Grasshopper		Insect	Orthoptera				
53	Field crickets		Insect	Orthoptera				
63	Locusts		Insect	Orthoptera				
39	Phylliidae		Insect	Phasmatodea	Phylliidae			Walking leaf insects
38	Phasmatodea		Insect	Phasmatodea				Stick insects
32	Plecoptera		Insect	Plecoptera				
30	Trichoptera		Insect	Trichoptera				
12	Insect		Insect					
9	Hymenoptera		Insecta	Hymenoptera				
50	Asian corn borer moth, Ostrinai furnacalis		Insecta	Lepidoptera	Crambidae	Ostrinia	Ostrinia furnacalis	Asian corn borer moth
56	Gypsy moths, Lymantriidae		Insecta	Lepidoptera	Erebidae			Gypsy moths

(continued)

(continued)

order in text	Index of taxa as they appear in the text	other taxonomic classes	Class	Order	Family	Genus	Species	Common name
51	Spilosoma niveus		Insecta	Lepidoptera	Arctiidae	Spilosoma	Spilosoma niveus	
140	Ostrinia nubilalis, Pyralidae		Insecta	Lepidoptera	Crambidae	Ostrinia	Ostrinia nubilalis	
52	Rhyparoides amurensis, Erebidae		Insecta	Lepidoptera	Erebidae	Rhyparioides	Rhyparoides amurensis	
116	Australian Granny's cloak, Speiredonia spectans, Noctuidae		Insecta	Lepidoptera	Erebidae	Speiredonia	Speiredonia spectans	Australian Granny's cloak
57	Geometrid moths		Insecta	Lepidoptera	Geometridae			Geometrid moths/Geometer moth
55	Winter moths, Geometridae		Insecta	Lepidoptera	Geometridae			Winter moths
49	Hepialids, Hepialus humuli		Insecta	Lepidoptera	Hepialidae	Hepialus	Hepialus humuli	
47	Hepialidae		Insecta	Lepidoptera	Hepialidae			
45	Lasiocampidae		Insecta	Lepidoptera	Lasiocampidae			
139	Pseudaletia unipuncta, Noctuidae		Insecta	Lepidoptera	Noctuidae	Mythimna	Pseudaletia unipuncta	
46	Nymphalidae		Insecta	Lepidoptera	Nymphalidae			
20	Rosettus leschenaulti	Megachiroptera	Mammalia	Chiroptera	Pteropodidae	Rosettus	Rosettus leschenaulti	Leschenault's rousette
18	Rousettus aegyptiacus	Megachiroptera	Mammalia	Chiroptera	Pteropodidae	Rosettus	Rousettus aegyptiacus	Egyptian fruit bat
19	Rousettus amplexicaudatus	Megachiroptera	Mammalia	Chiroptera	Pteropodidae	Rosettus	Rousettus amplexicaudatus	Geoffroy's rousette
17	Flying foxes	Megachiroptera	Mammalia	Chiroptera	Pteropodidae			
21	Old World fruit bats	Megachiroptera	Mammalia	Chiroptera	Pteropodidae			
131	Cross-banded treefrog, Smilisca sila		Mammalia	Anura	Hylidae	Smilisca	Smilisca sila	Cross-banded treefrog

(continued)

(continued)

order in text	Index of taxa as they appear in the text	other taxonomic classes	Class	Order	Family	Genus	Species	Common name
129	Tungara frog, Engystomops (=Physalaemus) pustulosus		Mammalia	Anura	Leptodactylidae	Engystomops	Engystomops (=Physalaemus) pustulosus	Tungara frog
130	Leptodactylus		Mammalia	Anura	Leptodactylidae	Leptodactylus		
6	Glossophaginae		Mammalia	Chiroptera	Glossophaginae			
113	Rhinonycterid Cloeotis percivali		Mammalia	Chiroptera	Hipposideridae	Cloeotis	Cloeotis percivali	Percival's trident bat
117	Dusky leaf-nosed bat, Hipposideros ater		Mammalia	Chiroptera	Hipposideridae	Hipposideros	Hipposideros ater	Dusky leaf-nosed bat
23	Hipposideridae		Mammalia	Chiroptera	Hipposideridae			
128	Cardioderma cor		Mammalia	Chiroptera	Megadermatidae	Cardioderma	Cardioderma cor	Heart-nosed bat
126	Megaderma lyra		Mammalia	Chiroptera	Megadermatidae	Megaderma	Megaderma lyra	Indian false vampire bat
132	Megaderma spasma		Mammalia	Chiroptera	Megadermatidae	Megaderma	Megaderma spasma	Lesser false vampire bat
3	Megaderma		Mammalia	Chiroptera	Megadermatidae	Megaderma		
123	Velvety free-tailed bat, Molossus molossus		Mammalia	Chiroptera	Molossidae	Molossus	Molossus molossus	Velvety free-tailed bat
105	Otomops martiensseni		Mammalia	Chiroptera	Molossidae	Otomops	Otomops martiensseni	Large-eared free-tailed bat
106	Tadarida aegyptiaca		Mammalia	Chiroptera	Molossidae	Tadarida	Tadarida aegyptiaca	Egyptian free-tailed bat
135	Tadarida teniotis, Molossid		Mammalia	Chiroptera	Molossidae	Tadarida	Tadarida teniotis	European free-tailed bat
136	Tardarida australis		Mammalia	Chiroptera	Molossidae	Tadarida	Tardarida australis	White-striped free-tailed bat
112	Mastiff bat		Mammalia	Chiroptera	Molossidae			Mastiff bat

(continued)

(continued)

order in text	Index of taxa as they appear in the text	other taxonomic classes	Class	Order	Family	Genus	Species	Common name
24	Pteronotus parnellii		Mammalia	Chiroptera	Mormoopidae	Pteronotus	Pteronotus parnellii	
124	Mormoopidae		Mammalia	Chiroptera	Mormoopidae			
133	Mystacina tuberculata		Mammalia	Chiroptera	Mystacinidae	Mystacina	Mystacina tuberculata	New Zealand lesser short-tailed bat
125	Nycteris grandis		Mammalia	Chiroptera	Nycteridae	Nycteris	Nycteris grandis	Large slit-faced bat
8	Tube-lipped nectar bat, Anoura fistulata		Mammalia	Chiroptera	Phyllostomidae	Anoura	Anoura fistulata	Tube-lipped nectar bat
78	Lophostoma silvicolum		Mammalia	Chiroptera	Phyllostomidae	Lophostoma	Lophostoma silvicolum	White-throated round-eared bat
68	Micronycteris hirsuta		Mammalia	Chiroptera	Phyllostomidae	Micronycteris	Micronycteris hirsuta	Hairy big-eared bat
67	Micronycteris microtis		Mammalia	Chiroptera	Phyllostomidae	Micronycteris	Micronycteris microtis	Common big-eared bat
138	Micronycteris microtus		Mammalia	Chiroptera	Phyllostomidae	Micronycteris	Micronycteris microtus	Common big-eared bat
2	Trachops		Mammalia	Chiroptera	Phyllostomidae	Trachops	Trachops cirrhosus	Fringe-lipped bat
127	Fringe-lipped bat, Trachops cirrhosus		Mammalia	Chiroptera	Phyllostomidae	Trachops	Trachops cirrhosus	Fringe-lipped bat
61	Rhinolophus ferremequinum		Mammalia	Chiroptera	Rhinolophidae	Rhinolophus	Rhinolophus ferremequinum	Greater horseshoe bat
1	Rhinolophus		Mammalia	Chiroptera	Rhinolophidae	Rhinolophus		
22	Rhinolophidae		Mammalia	Chiroptera	Rhinolophidae			
120	Barbastelle bat, Barbastella barbastellus		Mammalia	Chiroptera	Vespertilionidae	Barbastella	Barbastella barbastellus	Barbastelle bat
44	Eptesicus fuscus		Mammalia	Chiroptera	Vespertilionidae	Eptesicus	Eptesicus fuscus	Big brown bat

(continued)

(continued)

order in text	Index of taxa as they appear in the text	other taxonomic classes	Class	Order	Family	Genus	Species	Common name
48	Eptesicus nilssonii		Mammalia	Chiroptera	Vespertilionidae	*Eptesicus*	*Eptesicus nilssonii*	Northern bat
110	Spotted bat, Euderma maculatum, Vespertilionidae		Mammalia	Chiroptera	Vespertilionidae	*Euderma*	*Euderma maculatum*	Spotted bat
134	Euderma maculatum		Mammalia	Chiroptera	Vespertilionidae	*Euderma*	*Euderma maculatum*	Spotted bat
137	Lasiurus borealis		Mammalia	Chiroptera	Vespertilionidae	*Lasiurus*	*Lasiurus borealis*	Eastern red bat
122	Myotis blythii		Mammalia	Chiroptera	Vespertilionidae	*Myotis*	*Myotis blythii*	Lesser mouse-eared bat
4	Myotis		Mammalia	Chiroptera	Vespertilionidae	*Myotis*	*Myotis daubentonii*	
118	Myotis evotis		Mammalia	Chiroptera	Vespertilionidae	*Myotis*	*Myotis evotis*	Long-eared myotis
121	Myotis myotis		Mammalia	Chiroptera	Vespertilionidae	*Myotis*	*Myotis myotis*	Greater mouse-eared bat
90	Mouse-eared bat, Myotis myotis		Mammalia	Chiroptera	Vespertilionidae	*Myotis*	*Myotis myotis*	Mouse-eared bat
72	Myotis septentrionalis		Mammalia	Chiroptera	Vespertilionidae	*Myotis*	*Myotis septentrionalis*	Northern long-eared bat
119	Myotis septentrionalis		Mammalia	Chiroptera	Vespertilionidae	*Myotis*	*Myotis septentrionalis*	Northern long-eared bat
59	Nyctophilus geoffroyi		Mammalia	Chiroptera	Vespertilionidae	*Nyctophilus*	*Nyctophilus geoffroyi*	Lesser long-eared bat
41	Lizards		Reptilia	Squamata				
16	Vertebrates		Vertebrates					
25	Arthropods	Arthropoda						
7	Neotropical vine, Centropogon nigricans	Plant		Asterales	Campanulaceae	*Centropogon*	*Centropogon nigricans*	Neotropical vine
5	Balsa tree, Ochroma pyramidale	Plant		Malvales	Malvaceae	*Ochroma*	*Ochroma pyramidale*	Balsa tree